myBook+

Your online content to the book!

Online content that open up a digital dimension beyond a normal book. Depending on the topic, templates, information graphics, tutorials, videos or specially developed calculators – the myBook+ platform offers you all of this.

And it's that simple:

- Go to **https://mybookplus.de**, register, and enter your book code to get to your book
- **You'll find your individual book code at the end of the book**

We wish you lots of fun with myBook+ !

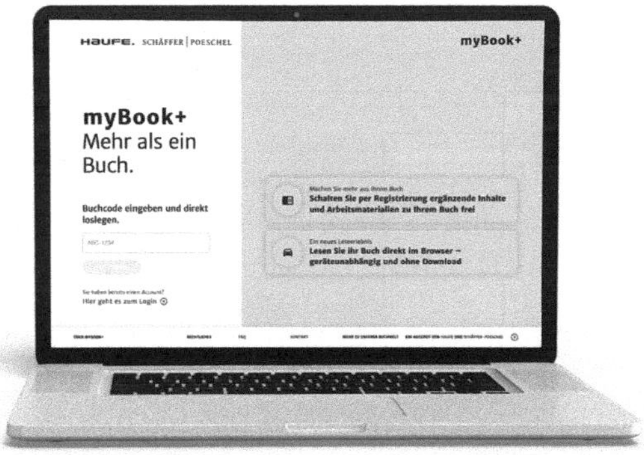

The Strategility Method

Kerstin Pichel/Thomas Haas/Bernhard Kruschitz

The Strategility Method

Agile strategy development: approach, tools and real-world examples

1st Edition

Schäffer-Poeschel Verlag Stuttgart

Bibliographical Information of the German National Library

The German National Library lists this publication in the German National Bibliography. Detailed bibliographical data can be accessed on the internet at http://dnb.dnb.de.

This edition is based on the translation of the 2nd edition of the book Kerstin Pichel/Thomas Haas/Bernhard Kruschitz, Strategility, 2nd edition 2024, also published by Schäffer-Poeschel Verlag Stuttgart.

Print:	ISBN 978-3-7910-6510-6	Order no. 12156-0001
ePub:	ISBN 978-3-7910-6511-3	Order no. 12156-0100
ePDF:	ISBN 978-3-7910-6512-0	Order no. 12156-0150

Kerstin Pichel/Thomas Haas/Bernhard Kruschitz
The Strategility Method
1st Edition, November 2024

Reinsburgstr. 27, 70178 Stuttgart, Germany
www.schaeffer-poeschel.de | service@schaeffer-poeschel.de

Image credits (cover): © Mo Büdinger
Illustrations: Mo Büdinger, www.buedinger.de
Product Management: Bettina Noé

Schäffer-Poeschel Verlag Stuttgart
A company of Haufe Group SE

Table of contents

Foreword

Take a guess: in the past year, you have heard of more than a dozen companies that are now also becoming agile, and in at most one case, this involves the »agilisation« of strategic decision-making. Yet the unlikely pairing of rapid agility and long-term strategy has a lot of potential.

For many employees, the term «strategy« conjures up associations such as lengthy management meetings and years-long implementation initiatives with little chance of success. The agile world, on the other hand, promises autonomy for teams and the ability to quickly respond to changes in the environment.

That's why we set out to learn from and with agile strategists. The following processes, tools, and examples are based on our experience, and have been developed as part of a research project called «Strategility«, funded by Innosuisse[1]. Since 2019, we have worked with more than 25 organisations to design, apply, reflect on, and further develop processes, tools, and roles of Agile Strategy Development, for example:

- in consulting projects with corporations such as Migros, with administrations such as the Swiss Federal Tax Administration, with the management of the Canton of Zurich, the Zurich University of the Arts, the town of Buch SG, in the energy industry such as EW Buchs, with SMEs, in IT such as VSHN AG and internezzo ag, with communication agencies such as Valencia AG, and many others
- in scientific work at companies such as SAP, Jobst Willers Engineering AG, and Scania Schweiz AG
- in continuing education courses in strategy moderation at the ZHAW, Zurich University of Applied Sciences, in continuing education courses in agile organisations at the FHNW, University of Applied Sciences, and Arts Northwestern Switzerland, in our public courses for Strategility Masters with participants from consulting, administration, and business and in company-specific training courses
- in the regular events of the Strategility Community, which now has over 100 members

These application experiences have confirmed that Agile Strategy Development improves the decision-making basis for development issues that are difficult to assess and can be combined with traditional strategy processes in administration and business.

This toolbox will give you an insight into how to systematically develop strategy content in an agile, nimble, and experimental way. It enables you to take advantage of agile methods and combine them with proven and familiar elements of conventional strate-

---

1 Innosuisse is the Swiss Agency for Innovation Promotion. It promotes application-orientated research projects. www.innosuisse.ch/.

gy development. Based on many examples from different industries and organisations and administrations of different sizes,[2] you will learn which approaches lead to success and which stumbling blocks you may encounter. Some of our application experiences are for example that:

- Organisations can and should adapt the processes and roles of the Strategility framework to their specific business situation. For example, in one international group all strategic hypotheses and tests were approved by the management, while in another the management was only involved in interpreting the outputs.
- The explicit formulation of strategic hypotheses and the open-minded evaluation of corresponding tests are unfamiliar to many decision-making bodies. Get expert support with these steps at the beginning, then you will succeed after three to five applications without further assistance.
- Review & Planning meetings should reflect information from all three Strategility processes in order to understand the strategic impact.

We recommend this toolbox to anyone involved in strategizing, regardless of their role in the organisation. We especially recommend it to those who want to organise a nimble and impact-oriented strategy process. We have created the role of the Strategility Master for them. If you have experience in Strategy Development and Implementation, you can use this toolbox to get started with Agile Strategy Development. A large number of templates and process descriptions – also available in the *Digital Extras* – enable you to get started immediately with Agile Strategy Development.

We would like to take this opportunity to thank our practice partners, the motivated Strategility Community, active users[3] in practice and trained Strategility Masters for their agile development work on this toolbox and the inspiring exchange of experiences. They have brought this toolbox to life:

Niklas Elser, Peppino Giaritta, Christian Hauri, Mike Keller, Lars Diener-Kimmich, Theo Knaus, Cem Kulac, Lars Lange, Thomas Landis, Claudia Meier Waldvogel, Andreas Meier, Alexander Mestre, Adrian Rüegsegger, Jeannine Spozio, Corinne Staub, Goran Vitkovic, Miriam Walther, André Woodtli, Andrea Müller, David Andrist, Dorothée Braun, Adrian Müller, Urs Villiger, Monika Gerber, Nils Arnold, Andreas Kreis, Adrian Staudenmaier, Leandro Gianetti, Vjekoslav Bosnjak, Florian Fasel, Marco Fretz, Daniel Bachmann, Claude Waeber, Lukas Rentsch, Franziska Gabriel, Nadja Heeb, Jane Wakefield, Thomas Molitor, and Nicole Kopp.

We would also like to thank our families and partners from the bottom of our hearts. They have kept us flexible during the development of this toolbox:

2 We use the terms ‹organisation and administration’ for companies in the private sector, NGOs, and public administration.

3 Gender-neutral language is used throughout this publication. The gender-neutral pronoun «they« has been used to ensure inclusivity. In instances where using «they« was not possible or would significantly impact readability, traditional gendered terms may be used.

Esther, Lena, Oliver; Lola, Ida, Markus; Eva, Jana, and Nina.

We look forward to continuing the lively exchange on Agile Strategy Development and therefore welcome your feedback and participation.

Agile Strategy Development and Implementation focuses the energy of your employees to the essentials and enables you to take overall control so that the strategy has the optimal impact on your organisation or administration. In doing so, you make a significant contribution to the future viability of your organisation.

CHAPTER 1

Agility for Your strategy

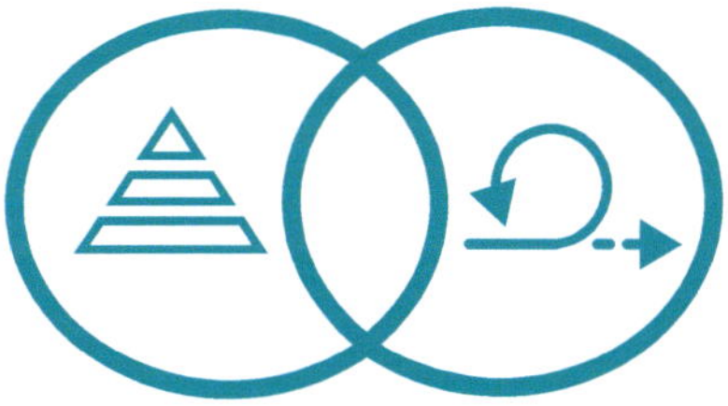

1 Agility for Your strategy

1.1 Your Use Case for Agile Strategy Development

Uncertainty in Strategic Decisions

Could your corporate strategy need an update? Does the future direction of your organisation seem static and outdated? At the same time, strategic impulses are constantly emerging and should not and cannot «only« be dealt with in your standardised strategy retreat. In short, does agility in strategy development appeal to you? Then welcome! You are in good company.

Agility is finding its way into many business development processes. You can use agile working methods to add value not only in your functional areas such as IT, product development, and HR, but also in your strategizing, because Agile Strategy Development – which we call Strategility – enables you to,

- create the best possible information base for uncertain decision-making situations;
- continuously record strategic information and integrate it into your decisions;
- increase your strategic response and implementation capability; and
- build an appreciative bridge between agile and strategic practitioners.

We have applied and reflected on agile strategizing with more than 25 organisations and administrations. We would like to start by outlining an impressive example of a use case and the added value of Agile Strategy Development.

Example

Agile Strategy Development Use Case

A company producing high-precision equipment for carbon analysis found itself in an initial situation where the global scientific market was largely saturated. In addition, new competitors were entering the market with new technologies. The company was in danger of losing growth, market power, and technological leadership. Changes in the customer segment and in product design were discussed. The company considered entering the new market of commercial production companies. The strategically important questions of what precision requirements and, above all, what operational requirements commercial customers would have were unresolved and the answers to these questions could not be determined analytically. The market was small and not described in industry reports, the product was new and customer response was unpredictable. However, the production of a carbon analyser took months, and the development and sales costs were in the five-figure range. A typical case for an agile approach to strategic decision-making.

Do you also find yourself in a situation characterised by uncertainty in which strategic decisions have to be made? Figure 1.1 provides you with a structure for your initial considerations – for a detailed description, see Chapter 2.2. Agile strategizing is particularly valuable if you do not know exactly which direction and strategy will lead your organisation to strategic success.

Fig. 1.1: Uncertainty as a driver of agile strategizing

VUCA as the New Normal

Uncertainty is becoming increasingly important as a framework for strategic decisions. Beginning in the 1980s, after the Cold War and the dissolution of the bloc powers, and continuing through megatrends such as globalisation, resource scarcity, digitalisation, and artificial intelligence, uncertainty has increased in relation to effective strategic decisions. We are currently experiencing the uncertain consequences of global supply chain disruptions due to the Covid-19 pandemic and the war in Ukraine. However, it is not only the economy that is facing enormous challenges, but also governments, administrations, and the healthcare sector. These are summarised by the term VUCA world.

Figure 1.2 describes the challenges of VUCA and the corresponding responses.

CHALLENGE	**VOLATILITY** Ambiguity and contradictoriness are increasing.	**UNCERTAINTY** Uncertainty and insecurity are increasing.	**COMPLEXITY** Influencing factors and interactions are becoming more powerful.	**AMBIVALENCE** Ambiguity and contradictoriness are increasing.
	V	**U**	**C**	**A**
REACTION	**VISION** Ensuring alignment with strategic orientation parameters.	**UNDERSTANDING** Understanding through involvement in development and through ongoing knowledge gains.	**CLARITY** Clarity about interactions through feedback of impact information. Clarity about responsibility through clear structures.	**AGILITY** Increase responsiveness to changes.

Fig. 1.2: VUCA – the challenge of time and the response

1.2 Mission, Vision, Strategy – Categorisation

What do we actually mean when we talk about strategy? The term «strategy« is often used, but rarely clearly defined. It is frequently used in very abstract terms as a «medium-term plan for the company's development«. This makes it difficult to answer the question: Does your organisation have a strategy? But rest assured: your organisation does have a strategy – even if it may not be consciously or fully defined. Ideas about specific strategy elements often exist in the minds of different people in the organisation and are added together through discussions and decisions to form a medium-term plan for the organisation's development.

Raising awareness of these strategy fragments and coordinating them with one another is worthwhile at the latest when the strategic development of your organisation is no longer progressing satisfactorily.

Strategy

A strategy is the definition of coordinated key activities. It defines the core areas of the organisation's internal commitment for the next one to four years. The following strategy elements should be defined:

- ***Playing fields/arenas:*** Which products and services does the organisation offer to which customer groups in which geographical markets? What are the organisation's core competencies and value chain?
- ***Differentiation:*** What advantages does the organisation use to position itself against competitors and customers?
- ***Vehicles:*** How does the organisation continue to develop itself?
- ***Timing:*** What is the speed and pace of strategic development?
- ***Economic logic:*** Which mode does an organisation choose to secure financial resources?

Strategic development often comes to a standstill if these focal points are not aligned with changes in the environment or in the company, or if they no longer fit together. For example, if the manufacturer of analytical instruments now focuses on quality management in the pharmaceutical industry, but continues to design its analytical instrument with complex analytical parameters that need to be adjusted individually, successful marketing is likely to be difficult. A coordinated strategy gives your organisation the strength to respond effectively to strategic challenges.

The five strategy elements outlined can be defined for three different organisational levels:

- At the group or company level, the strategy defines the business areas in which an organisation operates and which of these will be the focus over the next one to four years. This decision forms the basis for the allocation of resources and investments.

The intensity of the management role and the service activities of the company or group headquarters are also defined at this strategic level.

- Decisions on competitive activities are made at the business unit level, for example on specific customer groups, service offerings, and service activities in the relevant market environment. These decisions should be aligned with any overarching Group or overall corporate strategy.
- Functional strategies are defined at departmental level, such as marketing, purchasing or IT strategies. These strategies define the development priorities of areas within the organisation. They should also be competitive by focusing on the division's customers and their needs, differentiating them from external providers or managing their economic logic.

Strategies represent one of several Strategy Orientation parameters (see Figure 1.3). Today, strategies typically have a time horizon of two to four years. Strategic development priorities with a longer time horizon are described in more abstract orientation parameters such as the mission or vision. Strategic development priorities with a shorter time horizon are defined in more concrete guidelines such as project and action plans.

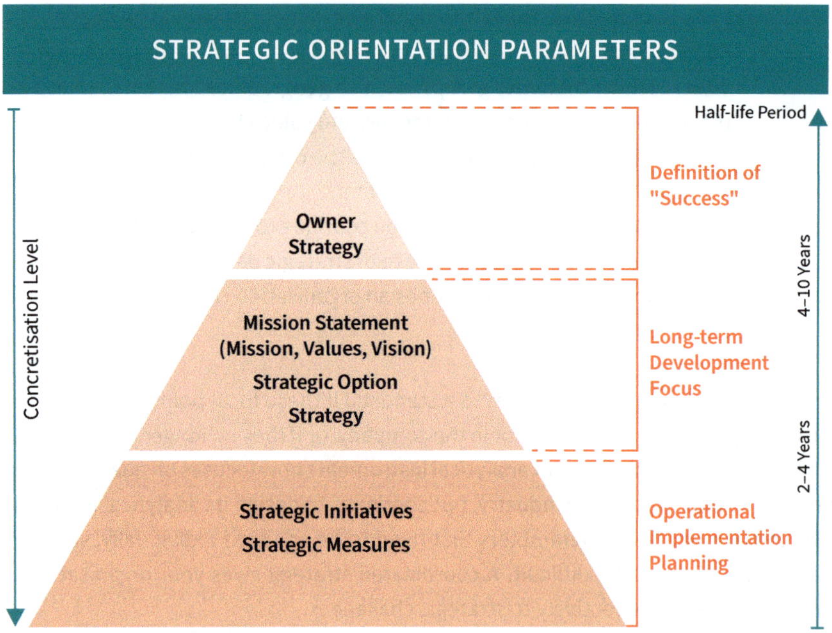

Fig. 1.3: Strategy Orientation parameters

1.3 Our Strategility Method for Your Strategizing

In this toolbox, we introduce the Strategility method to help you deal with uncertainty in strategic decision-making situations.

Agile Strategy Development

Agile Strategy Development – Strategility – is an iterative, evidence-based and impact-oriented method of strategizing. It involves the relevant knowledge holders in developing and implementing the strategy in collaborative teams with a high degree of ownership.

Strategility is a process with a continuous flow of information and insights between three main activities: Strategy Orientation with fundamental directional decisions, evidence-based Strategy Development and impact-oriented Strategy Implementation (see Figure 1.4).

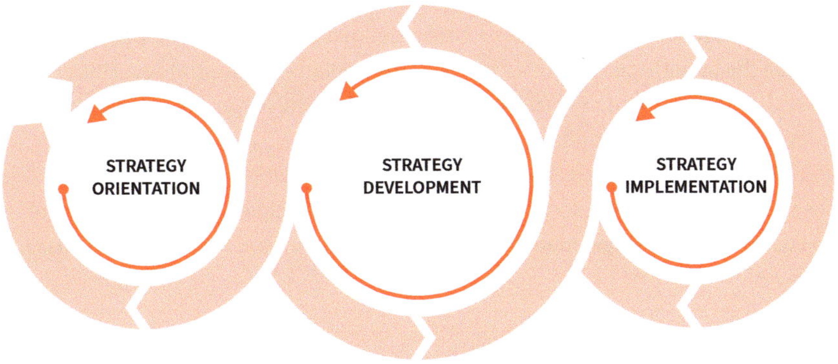

Fig. 1.4: Strategility process overview

This process is based on five principles that are consistently applied:

- **Alignment**
  Consistently base your strategic thinking and decisions on your organisation's strategic challenge and direction. You define this in the mission, values, vision, and strategic option (see Chapter 3). The strategy definition itself can be your link between the long-term orientation and the short-term operational activities. You can use the strategy definition as a learning-based dovetailing of the two development horizons, thus promoting a constant strategic orientation – an alignment.
- **Evidence**
  Make your decisions conscientiously, explicitly, and rationally, based on the best information you can gather. If necessary, you should also work with hypotheses and tests for Strategy Development (see Chapters 4 and 8) – because every decision is based on assumptions. If you make these explicit and discuss them, you can consistently analyse the impact of your Strategy Implementation and use them as evidence for strategy evaluation (see Chapter 5).
- **Iteration**
  Design an iterative process in which strategic insights are deliberately generated and continually used for the evaluation of current and past assumptions and decisions (see the process overview in Chapter 2, as well as details in Chapters 3, 4, 5, and 8). As

a result, strategic decisions remain rationally comprehensible while the initial implementation experiences are integrated.

- **Collaboration**
  Develop your basis for decision-making in a deliberate process with experts and decision-makers. This will enable you to gather as much information and interpretation as possible and to make the decisions that have been made or are still pending broadly comprehensible and actionable (see the explanations of the corresponding roles in Chapter 6). You can organise the necessary decision-making formats and processes in a company-specific way. The tasks, competencies, and responsibilities of the roles involved are then assigned according to your culture (see Strategility culture in Chapter 7).
- **Autonomy**
  Allow decision-making autonomy within defined guidelines to the people who will drive the hypotheses and tests forward and implement the strategy. This will save you from time-consuming approval processes (see the explanations on corresponding roles in Chapter 6 and on culture in Chapter 7).

These five principles serve to combine the different principles – or paradigms – of the two concepts of «strategic management« and «agility« in the Strategility process (see Figure 1.5).

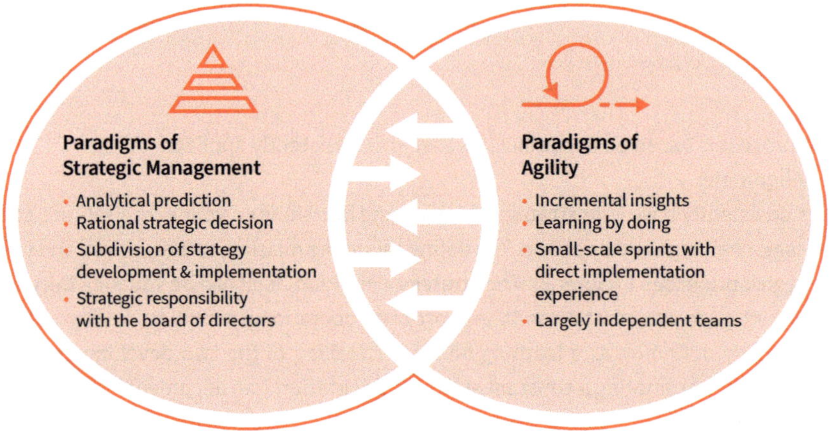

Fig. 1.5: Strategility combines paradigms from two concepts

The analytical approach of strategic management, with its alignment and cascading of responsibility for business development, is linked to the empirical approach of agility through the principles (see more detailed explanations in Chapter 7).

In the following chapters, we would like to explain the process design according to these principles as well as the content, sequence, roles, and tools of the Strategility process in more detail.

1.4 Strategility Flow and Framework – Overview

Do you want to accelerate the strategic development of your organisation? Would you like to be much further ahead, but the strategic alignment always takes longer than desired? At the same time, do you want to keep your strategic options open and be able to react immediately if necessary? The Strategility framework supports continuous strategizing and thus contributes to your ability to deal with change. The framework promotes:

- **Effectiveness** – derive appropriate strategies thanks to a multidimensional basis for decision-making
- **Efficiency** – achieve insights and impact faster
- **Flexibility** – mitigate risks, change plans, change direction

The Strategility framework (see Figure 1.6) includes the following elements:

1. The **Strategility flow** represents the core of the framework. It is used to methodically manage continuous strategizing with the three sub-processes of Strategy Orientation, Strategy Development, and Strategy Implementation.
2. **Events** for planning, controlling, and reviewing Strategy Development and its outputs.
3. **Outputs**
   a) Strategic challenge
   b) Mission statement
   c) Strategic option
   d) Key output strategy with strategy canvas
   e) Business case and strategy roadmap
4. **Roles** with specific tasks and defined responsibilities
5. **Tools** for strategizing, including:
   a) Strategy Orientation
   b) Strategic challenge canvas
   c) Strategility Board
   d) Risk tower
   e) Hypothesis and test canvas
   f) Strategy canvas
   g) Project canvas
   h) Sequences of events
6. **Principles**

The Strategility framework is based on the paradigms of agility and strategic management. The combination leads to:

- impact- and risk-oriented decision-making through comprehensive strategic decision information
- action-relevant decisions through implementation-oriented strategic hypotheses and tests
- fast responsiveness through a constant flow of strategic information

The following case study illustrates agile strategizing:

- Combining established and agile methods
- Iterative way of working in interdisciplinary teams
- Learning culture and evidence-based decisions

The case study outlines the Strategy Development over the course of two years, from the preparatory analysis of the Strategy Orientation (see Chapter 3) through five iterations of Strategy Development (see Chapter 4) to partial Strategy Implementation (see Chapter 5). The tasks of the individual roles are described in Chapter 6 and 7 addresses the culture. Hypothesis-driven work and experiments are described in Chapter 8.

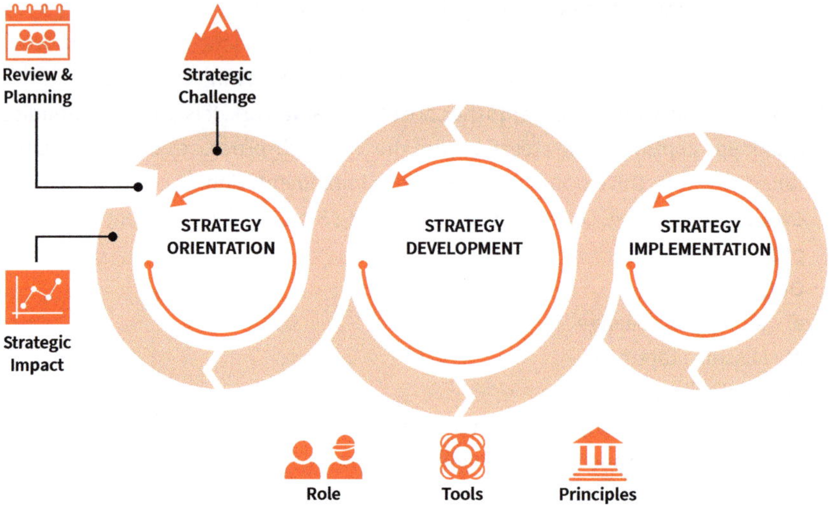

Fig. 1.6: The Strategility Framework

> **Example**
> **Department Store – Challenge: Decline of Bricks-and-mortar Retailing**
>
> **Strategic challenge:**
> Department stores are under pressure in stationary retail due to a variety of changes and influences, such as the trend towards online shopping and the minimalist lifestyle. A Swiss department store in the upmarket segment recognised the strategic challenge of securing its existence for the coming years despite these developments and decided to respond to declining sales figures. Based on analyses and studies of consumer behaviour, the management decided to embrace the sharing economy trend: Borrowing instead of buying – a radical idea for a department store.
> This analysis of trends, market developments, consumer behaviour, and innovations in the retail sector around the world took around a year to complete.

Other threats to bricks-and-mortar retail and trends, such as the digitalisation of the shopping experience through digital displays, AI-based shopping assistants, etc., were put on the back burner.

Strategic option:

Product development. The central assumption is that consumers prefer to use goods over time rather than own them.

Initial strategy:

The rental of parts of the product range to existing customers in the existing stores.

No reliable strategy could be formulated using established conceptual-analytical methods of Strategy Development. Questions about the product range, customer segment, pricing models, processes, logistics, and prospects for success remained unanswered due to a lack of data and benchmarks. The strategic option (see Chapter 3.3) and the strategy were developed and validated iteratively over the course of a year. The work was designed to maximise knowledge gain. Strategy Orientation, Strategy Development, and Strategy Implementation were driven forward simultaneously. The following was carried out and discovered in iterations of two to three months each:

1. **The field test**
   Who is interested in «borrowing instead of buying«?
   In a shop, all non-food items could be borrowed instead of bought for two weeks – including underwear and toothpaste. To keep the process simple, borrowed goods were purchased at a normal shop checkout by project staff using a dedicated credit card and managed in Excel. The returns process consisted of a single point of contact for customer service.
   The result: virtually nothing was borrowed. It was recognised that the more conservative and affluent clientele wanted to own goods and could afford to do so.
2. **The pop-up store**
   Do young, urban people behave differently?
   For the follow-up experiment, a pop-up store with a specially created brand was operated in a trendy neighbourhood for two weeks. In addition to a small range of clothing, accessories, and furnishings, every item in the range could be rented from the pop-up store.
   The result: Young women borrow expensive clothes.
3. **The fake online shops**
   What is the right rental range?
   In order to find the right rental range, several «fake online shops« were placed on the Internet, each with a very specific product range. The shops were advertised and their success measured using analytics. They were labelled as «fake online shops« because the ordering process was done by email and no other shop functionality was implemented.
   The result: Handbags performed best.

4. **The fake process**
   Can you earn money by renting?
   In the following iteration, the marketing of the rental service focused on handbags. The term «fake process« was used because each order was regularly purchased by a team member at a store checkout and delivered to the post office.
   The result: In the foreseeable future, rentals would exceed sales and could be operated profitably.
5. **The strategic decision**
   The results of the iterations, combined with market research and surveys, were used to formulate a sharing economy strategy with validated strategy elements.

1.5 The Tools

The Strategility framework framework provides you with a wide range of tools for strategizing. In addition to the images and illustrations in the book, you can download and use templates directly from our online materials, the *Digital Extras.* The tools include

1. **Orientation and controlling**
   - Strategic challenge
   - Strategility Board
   - Strategility impact indicators
   - Assumptions and hypotheses
   - Tests and experiments, test results
   - Knowledge-driven prioritisation (risk reduction)
2. **Strategy Orientation**
   - SWOT analysis
   - Strategy Orientation parameters incl. mission statement with mission, vision, and values
   - Checklist of framework conditions for strategic options
   - Strategic benchmarks for the administration
   - Checklist of framework conditions for strategic options for the administration
   - SWOT3
   - Evaluation of strategic options
   - Canvas strategic challenge
   - Canvas risk tower
3. **Strategy Development**
   - Strategy canvas with the strategy elements
   - Assumptions canvas
   - Hypothesis canvas
   - Test canvas
4. **Strategy Implementation**
   - Project canvas

5. **Roles**
   - Strategility roles and filling
   - Process steps and roles with responsibility and participation
   - Role descriptions
6. **Corporate culture**
   - Self-assessment of corporate culture according to Schneider (1994)

1.6 Getting Started with Agile Strategizing

Have you come to the realisation that an agile Strategy Development meets your strategic decision situation and your needs? Then bring together strategic thinkers and agile contributors and appoint someone to design and moderate the strategizing process – we call them the Strategility Masters. You can find out more about this in Chapter 6.

The Strategility flow turns strategizing into a continuous process. We call these Strategility iterations, the duration of which depends on the characteristics of the organisation or administration and its challenges (see Chapter 7). The following procedure has proved to be useful for starting the first iteration:

- Strategility Master: If you have no experience with Strategility, we recommend that you attend a course and/or be accompanied by a Strategility consultant.
- Create a process design: What are the steps and who is involved? The cultural analysis serves as a basis (see Chapter 7).
- Carry out a strategy kick-off: As an introduction to the process, principles, roles, and working methods, the Strategility Master presents and establishes the working methods.

Event
Strategility Kick-off
Objective: Common understanding of the Strategility objective, approach, and organisation of roles
Participants: Strategility Masters, Enablers, potential Strategy Owners, potential Hypothesis Owners, Strategy Exploration Team
Work steps:

- Presentation and inquiry about prior knowledge of Strategility (prerequisite for strategizing), reference to planned introductory event if applicable
- Explanation of the objectives and expected benefits of agile strategizing
- Explanation of the content-related framework conditions and the strategic guidelines: Owner's strategy, mission, vision (see Chapter 3)
- Explanation of the Strategility approach and explanation of the implementation of the recurring Review & Planning events (see Chapter 2)
- Roles and role allocation; clarification of tasks, competencies, responsibilities, and mutual expectations of the role holders (see Chapter 6)

- Identify success factors for joint agile strategizing; possible questions could be: How consistently do we want to deal with unexpected and undesirable test results? How do we ensure that decision-making autonomy is not detrimental to the organisation as a whole? How do we balance the need to be agile with the need to be strategic?
- Define further procedure for agile strategizing

Result: Roles and work process are clarified, collaboration is initialised, and success factors are recorded.

Next steps: Develop the strategic challenge and orientation (see Chapter 3); derive risks from success factors and include risk mitigation measures in the Strategility Board.

1.7 The Potential of Strategility

Strategility has shown the following potential in our partner companies:

- They gain effectiveness in their strategic decisions because they recognise at an early stage which strategy is effective and which is not through continuous knowledge acquisition and evaluation.
- They gain effectiveness in Strategy Implementation because they integrate the potential of many knowledge holders through co-creation and interdisciplinary cooperation and align it with the strategic challenge and the strategic options.
- They gain efficiency because a clear focus on long-term guidelines such as mission, vision, and strategic option enables autonomous and fast-acting teams to test strategy elements in the interest of your organisation.
- They gain flexibility because they reflect on the effectiveness of your strategic decisions in regular review and planning meetings.

We would be proud to see you start with agile Strategy Development. With this toolbox, we are happy to share our knowledge and recommendations with you.

CHAPTER 2

Strategility Method and Flow

2 Strategility Method and Flow

2.1 The Strategility Flow

The Strategility flow ensures that all strategizing is methodologically sound (see Figure 2.1). The flow is divided into three sub-processes:

1. **Strategy Orientation** creates orientation with a mission statement and strategic option.
2. **Strategy Development** formulates elements of a strategy, tests its assumptions with hypotheses, tests, and experiments and derives a validated overall strategy.
3. **Strategy Implementation** puts the strategy into practice through initiatives, plans them, monitors their output and impact, and ensures continuous feedback into the Strategy Development.

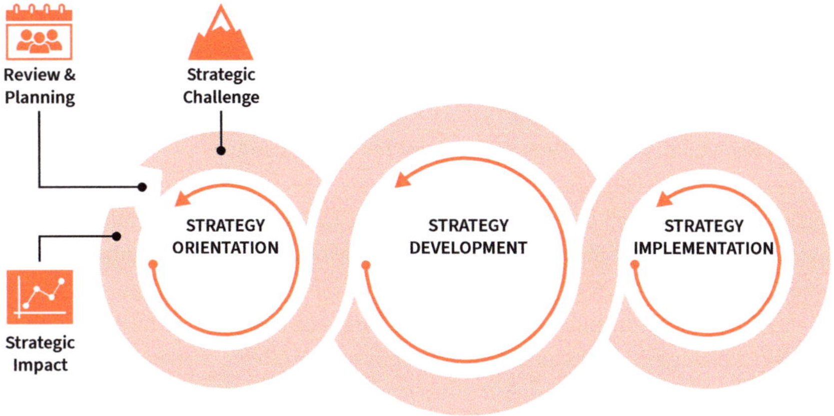

Fig. 2.1: The Strategility Flow at a glance

The main task of the Strategility flow is to master the strategic challenge. The steering of strategizing is guided by its impact on the strategic challenge. The explicit formulation of a strategic challenge is an important element of the method. The Strategility flow is based on a step-by-step approach and is divided into iterations, typically lasting several months. Planning takes place at the beginning of an iteration and is concluded with a review at the end. As the iterations form a sequence, the review of the previous iteration and the planning of the subsequent iteration are combined into one event (see Chapter 2.7.1).

Within an iteration, the three sub-processes are active at the same time. This means that Strategy Orientation, Strategy Development, and Strategy Implementation are interlinked (see also Chapter 2.7 on management). The common pace ensures synchronisation: clear alignment, evaluation of findings, joint learning. This leads to the special feature of agile strategizing: insights on outputs and impacts from all stages flow back into the upstream process steps. In this way, decisions made upstream remain flexible to change:

- **Decision flow**
  Strategy Orientation → Strategy Development → Strategy Implementation
- **Feedback of results**
  Strategy Implementation → Strategy Development → Strategy Orientation

The main outputs of the overall process in the Strategility flow are:

- the strategic challenge with impact indicators (see Chapter 2.3)
- the process design for the needs-based Strategility flow
- the Strategy Orientation parameters such as vision and strategic option (see Chapter 3)
- the formulated strategy (see Chapter 4)
- the implemented strategic initiatives (see Chapter 5)

By working in and with the Strategility flow, the organisation develops:

- a common Strategy Orientation (see Chapter 2.7.1)
- a learning culture and adaptability
- strategic impact at all levels

The Strategic Challenge as the Starting Point for Strategizing

Strategizing begins with the formulation of the strategic challenge (see Figure 2.2): Which strategic drivers – problems, opportunities, changes – should be addressed and overcome by the upcoming strategizing in order to fulfil the overriding interests of the owners? What strategic challenge needs to be overcome so that the organisation's purpose can continue to be fulfilled in the future? Strategizing is steered by mastering the strategic challenge; it is concluded by overcoming the strategic challenge.

Is strategizing ever finished in a complex and dynamic environment? Probably not. However, the strategic challenge will change over time as the strategizing is developed – triggered, for example, by changing external circumstances or newly required internal capabilities. When this happens, the drivers of strategizing shift or change – one challenge is met and the organisation can already prepare for the next!

This shift was clearly evident in the years 2019 to 2022 in the example of resource utilisation and sourcing. Today, every company has to ask itself questions about production locations, resource use, and energy mix. While the climate and health movements were strong drivers in the past, the Covid-19 pandemic pushed them into the background and focused attention on dependencies on global supply chains. With the war in Ukraine, sanctions against Russia brought back the issues of energy supply and the sourcing of raw materials and semi-finished products, as well as some of the concerns of the climate movement. At the time of writing, the long-term impact is difficult to assess. Will longer delivery times, higher commodity prices, and more unreliable supply chains become the norm? How will suppliers, competitors, customers, the public, and politicians react – and how will you?

Practical Tip

Flexible Strategility Flow

Strategizing with Strategility begins with process design and the formulation of the strategic challenge. The ideal flow results from the strategic challenge. The flow starts either in a clearly guided manner with Strategy Orientation or emergently, from an inner drive from the operational business, the implementation or from the development process.

The Strategility method and the Strategility flow can be used to start at any point, depending on the initial situation.

2.2 Storytelling in Strategizing

The aim of the Strategility Flow is to achieve maximum strategic impact. Strategic impact is achieved through sustainable, feasible, and viable solutions for strategic challenges. Therefore, management in the Strategility flow is focused on maximising impact rather than controlling a rigid sequence of activities. Agile strategizing is guided by three management principles:

- Focus on the strategic challenge
- Minimising risks by gaining certainty (see Chapter 2.4)
- Adding value step by step (see Chapter 2.5)

Focus on the Strategic Challenge

The strategic challenge is the motivation for strategizing. Why are you making the effort? What problem will be solved, in what way will your organisation be better off when the strategizing has been successfully completed? The function of strategy is to enable the organisation to fulfil its purpose adequately. If the organisation is not, or is no longer, able to fulfil this purpose in the long term, it is strategically challenged: for example, without an online service, a bank today can only serve a diminishing number of customers. The success of a strategy is measured by how well it meets this challenge. The Strategility framework for agile strategizing is particularly suited for the following strategic challenges:

- **Rapid changes in the outside world:** The VUCA world (see Chapter 1.1) is characterised by rapid changes in the outside world that confront organisations with important strategic decisions. The resulting challenge is exacerbated by megatrends such as digitalisation, demographic change, globalisation, and its antithesis, neo-ecology. However, «fast« is not an absolute value. Fast or too fast means that your organisation's established processes and structures are not able to make and implement decisions in a timely manner.
- **Complexity of the strategic option:** The strategy or strategic environment for the chosen option is too complex for proven, plan-driven strategizing. Complex is, for example, linking a department store's customer groups with a new rental offer and the logistical handling of the new rental offer. Each decision depends on the others and can hardly be outlined in advance. In the context of diversification, new ser-

vices have to be offered in new markets to new customers using new technologies. The interdependencies are nontransparent. Planning is difficult, and plans are too imprecise, too risky.

If you are aware of the strategic challenges facing your organisation, you will also – consciously or unconsciously – make assumptions about developments and effective solutions (see Chapter 2.4). The added value of strategizing lies in the creation of strategic planning certainty and the gradual mastery of challenges.

Strategility Storytelling Canvas for the Strategic Challenge

Established strategizing provides a wealth of proven tools for describing the strategic starting position. For example (see Chapter 3.4): SWOT, environment analysis PESTEL, 5-Forces analysis, competitor and customer analysis, and their summary in the SWOT presentation. Development can take place through a combination of workshops, analysis, and experimentation. It is guided by the storytelling canvas (see Figure 2.2):

1. The **questions** in the five areas guide the development of the strategic challenge.
2. The **strategy story** defines the cornerstones of storytelling.
3. The definition of **observables** sharpens the understanding of impact orientation.

Storytelling turns abstract strategizing into a concrete, tangible story. Without communicating the strategy, the workforce cannot be engaged. Without an engaged workforce, there is no strength in execution, nor can other groups add value to strategizing.

	Starting position	Driver	Consequences	Obstacles	Success
Question	Where are we coming from? What were our strengths? Where are we today?	What change is driving us? What doesn't work anymore?	What is our orientation, if we don't act?	Which uncertainties stand in our way? Where are the dangers lurking?	The challenge has been overcome. How does the success look?
Strategy story					
Observables					

Fig. 2.2: Using the storytelling canvas to meet the strategic challenge

The following anonymised strategy story was created in a customer project:

1. The Migration Office makes an important contribution to the development and security of the country. The specialists and employees have been able to cope with the constantly increasing demands. Progress in digitisation has been successfully implemented over the last 20 years.
2. The number of applications and the complexity of the cases will continue to increase, although the number of authorised staff will remain the same. Core system technology is disappearing from the market. Appropriate IT expertise is becoming increasingly difficult to find. The operation of the systems is not secure in the long term.
3. Without countermeasures, the office will no longer be able to fulfil its core tasks in the foreseeable future. Dissatisfied employees and increasing staff turnover will lead to a decline in the quality of service. Its reputation with the public and partners will decline.
4. With a new core system, the advances in digitisation can help to design processes and structures that can cope with increasing volume and complexity with a given number of staff.
5. Better working tools will promote employee satisfaction and the quality of service. The office remains a reliable partner for other offices, authorities, and municipalities and deserves the trust of the population.

Good storytelling is characterised by ease of understanding, comprehensibility, personal involvement, realistic statements, and authentic language. What story are you telling?

2.3 Strategility Impact Indicators

We measure the success of Strategy Development by the strategic impact, i.e. the mastery of the strategic challenge. We measure this impact using impact indicators. In practice, measurements and indicators often trigger resistance and fear. Observables can therefore be more accessible than indicators. An observable describes a shared understanding of a certain state, for example the drivers, the challenge or the strategic success. A shared understanding of the observation of conditions can be used to create a shared picture of the state and progress.

For some organisations, observables are good enough for the process. Others derive additional tangible, measurable indicators. In the example of the Migration Office, observables include service quality, satisfied employees, reliability as a partner, and reputation. These observables describe an effect and are ongoing. Tracking means that measures will have a delayed effect at some point in the future.

The tangible and measurable impact indicators create transparency about the success of strategizing. This is a once the strategic challenge has been overcome:

- The threat has been averted.

- The strategic option unfolds its impact.
- The key elements of the strategy have been implemented.

It is also advisable to keep an eye on the operating business with key performance indicators: turnover, margins, and capacity utilisation.

The indicators are continuously updated together with the results and interpreted at the end of each iteration at the latest. This means that a change in the indicators can be observed within an iteration. For example, in the department stores, customer frequency or sales are measured daily.

The impact indicators are used to guide strategizing: prioritise the next iterations accordingly and decide on specific measures and next steps.

Impact Indicators for Drivers and Effects

The impact of strategizing can be measured by indicators of the drivers. These must diminish over time:

- the threat of the strategic challenge
- the uncertainty of the assumptions

The success of strategizing can be measured by impact indicators. These must increase over time:

- impact on business performance such as sales, EBIT or operating profit as a leading indicator
- impact on orientation such as market positioning or shareholder value as a lagging indicator

The impact indicators for drivers and effects are presented in a four-field matrix for management purposes (see Figure 2.3). Successful strategizing leads to a decrease in the influence of the drivers on the company. The desired effects increase. The impact indicators serve as a snapshot and as a time series for managing Strategy Development.

The impact indicators represent the strategic challenge as a number and a statistical series. Even if some indicators, such as the company value, do not change or change only insignificantly during strategizing, this overall view of the strategic challenge is particularly valuable for managing Strategy Development: the impact indicators prevent the strategy team from getting lost in the details.

To be successful in the consumer goods rental business, for example, you need to gain a foothold in the market. The products and services must meet a need there to be well received. Attractiveness, demand and interest can be observed, measured, and evaluated. They are leading indicators of the potential success of strategizing. Lagging indicators are, for example, a strong market position or the development of the company's value. If the leading indicators do not show success, the lagging indicators will never materialise.

You develop the concrete outputs of Strategy Development and Implementation in iterations. It is uncertain whether the desired effect will be achieved. The impact indicators help you to find out whether your planned strategy is actually having a positive effect and responding to the strategic challenge. Start by focusing on the indicators that help you to assess momentous assumptions. If these point in an undesirable direction, all other assumptions and indicators become obsolete.

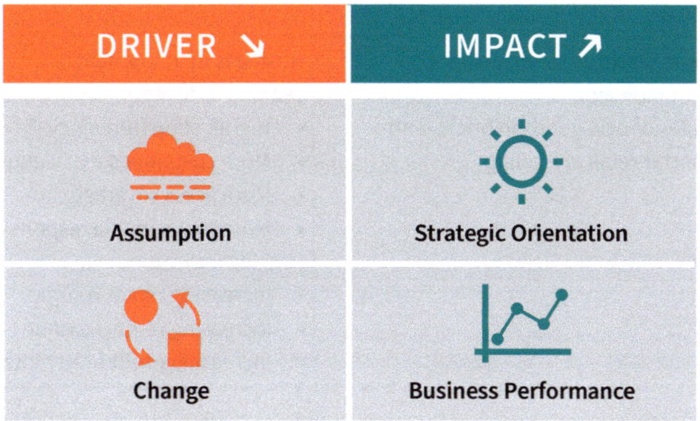

Fig. 2.3: Strategility impact indicators

> **Example**
> **Impact Indicators Product Development: Rental in the Department Store**
> Table 2.1 shows an example of the choice of impact indicators for the effects and drivers of the strategic option «Product development of our department store on the way to becoming a rental store«.
> **Change:**
> If consumer habits really do change, the industry's overall online and bricks-and-mortar sales should decline.
> **Acceptance:**
> The trend towards the sharing economy should be reflected in other areas: further rental offers for other consumer goods or the increasing popularity of existing offers (e.g. car sharing, car pooling, etc.).
> **Business development:**
> The success of the offer can be assessed in advance based on the response, such as customer frequency or enquiries. Rental volumes, turnover, and profitability react slowly or subsequently.
> **Strategy Orientation:**
> When the rental business reaches the profit zone, it delivers the desired business development. The profit zone and sustainable development are typically delayed, usually taking effect long after strategizing has been completed.

Driver: Rent instead of buy	Effect: Positive business development
Assumption: Consumers prefer to rent rather than buy Leading indicators: • New rental offers for consumer goods in various sectors • Second-hand trade on the rise • Enquiries from suppliers about rental offers	**Strategy Orientation:** Rental business compensates for losses in bricks-and-mortar retail Lagging indicators of success: • Reaching the profit zone • Stable sales growth • Stable development of operating costs
Change: Trade loses importance Current indicator: • Sales of new goods in bricks-and-mortar retail are falling	**Business development** Leading indicators – resonance: • Higher shop frequency • More customer service enquiries • More press coverage • More social media response Lagging indicators: • Increasing rental volume • Increasing rental revenue • Increasing profit from rental

Table 2.1: Drivers and impacts of the strategic option «Borrowing instead of buying market development«

2.4 Risk Orientation: Creating Certainty

Uncertainty paralyses decision-making and delays your initiatives. Correcting wrong decisions with serious consequences wastes resources and leads to a loss of time. You can counteract this with a learning culture based on the principles of «fail early, learn quickly«, as shown in Figure 2.5. To achieve this, plan work *in a risk-oriented manner*:

1. Create certainty to make evidence-based decisions : Make assumptions transparent and consistently align strategizing with the most momentous assumptions. For example, in the evolution from a department store to a rental store, this could be the assumption that the rental of kitchen appliances and jewellery will appeal to existing or new, interested customers.
2. Classically proven analytical methods (e.g. market analysis) are supplemented by empirical methods in order to gain knowledge: Hypothesis-based techniques such as experiments and tests (e.g. non-existent rental offers) to refine lengthy, costly ideas with serious consequences quickly, simply, and cost-effectively. In the case of pawnbroking, product categories, loan periods, and logistical requirements can be tested before a costly logistics centre is built.

You can avoid the cost of mistakes with serious consequences, accelerate strategizing and create confidence in decision-making.

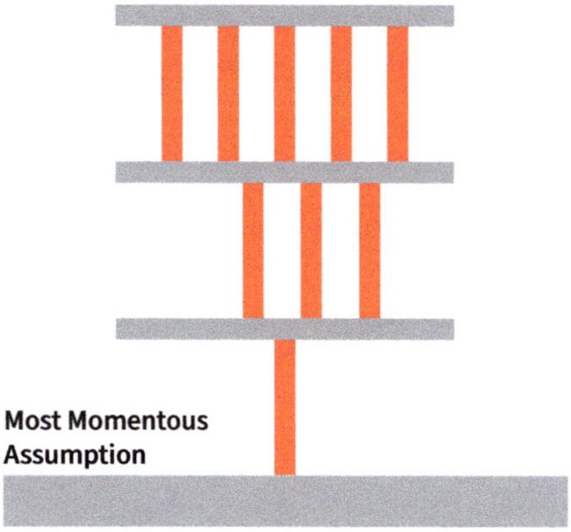

Fig. 2.4: Finding the most momentous assumptions with the risk tower

The risk tower (see Figure 2.4) helps to uncover hidden assumptions and unspoken beliefs. In the tower, the columns represent the risk: If a column falls, the part of the tower above it collapses. If one of the lowest columns falls, the consequences are most serious. If the top columns fall, the consequences are less severe or more tolerable.

The canvas can help you as a manager to reflect on assumptions. The canvas has also been used in workshops to draw on the experiences and views of many people and to bring hidden assumptions to light.

The risk tower not only visualises assumptions, but also ranks them according to urgency. This assessment is used to prioritise the work to be done in the «Review & Planning« iteration (see Chapter 2.7.1). Chapter 8.4 provides a workshop format that can be used to uncover and prioritise hidden assumptions using the canvas.

The most momentous assumptions with high uncertainty and high error costs are prioritised in strategizing.

Figure 2.5 shows the typical progression of increasing certainty and decreasing error costs in market and product development, from validating the idea to verifying the approach and scaling the solution. In the department store example these are

1. **Validate the idea:** Does «rent instead of buying« resonate? With whom?
2. **Verify the approach:** How does the customer experience need to be designed for renting? This includes the product range, conditions, and processes.
3. **Scale the solution:** Sustainable implementation of pilot organisations and processes

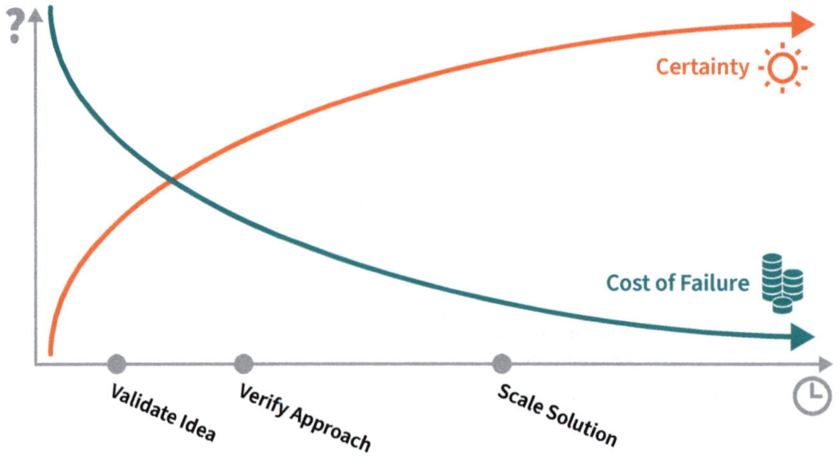

Fig. 2.5: Increase certainty and reduce the costs of failure

Example

Expansion of OVS from Italy to Switzerland

The fashion house OVS is very successful in Italy with fashion for the young and the young at heart. OVS has been less successful in expanding into the DACH region.

In Switzerland, OVS took over more than 100 stores of the Charles Vögele fashion house in 2016, remodelled them and converted the assortment to its own products. Sales collapsed and the stores had to be closed.

In Switzerland, Charles Vögele served a rather conservative, older clientele. The Italian, youth-oriented range in smaller sizes did not appeal to regular customers. Regular customers turned away from OVS shops. The OVS brand was not well known to the target group, the store network was too large and there were too few new customers.

The development of a store network is a typical result of Strategy Implementation. As a leading indicator, the store network shows progress in strategizing. The assumption that the market for OVS fashion could be developed from the branch network of Charles Vögele with its existing regular customers proved to be a serious misconception. There was no correlation between the result (store network) and the effect (serving the Swiss market). You can find out more about the hypothesis-based approach in Chapter 8.3.

2.5 Working in Iterations: Step-by-step Procedure

Lengthy initiatives make it difficult to change direction. Breaking such initiatives down into individual stages or steps has many advantages. For example, the transparency of the individual stages and steps on the Strategility Board makes it easier to monitor the progress of strategizing. In order for a change of direction to take place after individual Strategility iterations, the stage must be self-contained. It must bring concrete added value. And: they reduce risks, the costs of momentous assumptions and decisions.

In the development of tangible products, such stages are typically called prototypes, minimum viable products (MVP), trials, pilots, early access or beta testing. Feedback from users can be used to improve the unfinished product. The intended added value of each stage is derived from the findings, corrections, and optimisations.

In abstract strategizing, the added value of a stage is the planning certainty for the following steps: Assumptions such as customer interest become facts such as concrete enquiries or reservations for future orders. To gain agility, you should prioritise strategizing according to the greatest value added in the shortest time (see Figure 2.6). In economics, this procedure is called «prioritisation according to diminishing marginal utility«. In product development this is referred to as the cost of delay (cf. Reinertsen 2002). Prioritisation is based on value added, weighted by the duration of the task (Weighted Shortest Job First – WSJF).

Example

Product Development Carbon Analyser

In the example of the product development of a new carbon analyser from Chapter 1.1, there are both market and technical issues to be addressed. Which questions would you address first? Which questions will add the most value in the shortest time?

Table 2.2 shows examples of three of the many strategy tasks required to develop the new carbon analyser from Chapter 1.1. For the three tasks, the team estimates the relative added value of the output from small to large and the duration of the work from small to very large. For both estimates – added value and duration – the team uses a scale from 1 to 5. The higher the calculated WSJF value (relative value divided by relative duration), the lower the cost of delay. The prioritisation of tasks is derived from the WSJF value.

In the table, the market-oriented questions have a higher WSJF value and should be tackled first from an economic point of view. This leads to the following iterations:

1. Clarify the requirements for a new analyser with customer surveys
2. If there is a need, determine the specific need with a landing page
3. Develop the new analytics module

Task in strategizing	Relative value	Relative duration	WSJF value	Priority
Technology: Developing a new analysis module	4	5	0.6	3
Market: Customer survey on requirements for the new analyser	2	1	2	1
Market: Landing page with pre-order of the new analyser	1	1	1	2

Table 2.2: Prioritising strategizing with WSJF

Chapter 8.2 describes a supplementary workshop format that can be used to address the greatest uncertainties in the shortest possible time.

Practical Tip

Planning Poker

Estimating abstract benefits such as gained insights and contract costs can be very time-consuming. Much simpler, quicker, and proven in practice is the relative estimation of added value and duration.

The simplest form of relative estimation is to plot the tasks to be prioritised on two axes: one axis by value added, the other by duration. Similar to a risk matrix, the high priority tasks are in the high added value and low duration quadrant. With good facilitation, 30 tasks can be prioritised in an hour using sticky notes and a pin board.

Another form of relative estimation is planning poker. Here tasks are assigned to predefined orders of magnitude 1, 2, 3, 5, 8, 13, 21 of the Fibonacci series. The highest value added or greatest effort is assigned a 21, the lowest value or shortest duration is assigned a 1. The WSJF value is then calculated from the two values.

Facilitation tip: Form the series tasks on the two axes without fixed scales and assign the values of the orders of magnitude to the series formed.

Fig. 2.6: Added value in strategizing: the greatest gain in insights in the shortest possible time

2.6 Collaboration: Participating vs. Co-Deciding

In a dynamic and complex environment, experience has shown that interdisciplinary collaboration in self-organised teams is superior to the traditional, plan-driven, and division of labour organisation, because specialist and decision-making expertise can be pooled, decisions can be made quickly and coordination efforts can be reduced. Self-organised work is motivating. The interaction between the roles is just as important as the fulfilment of the individual roles. Constructive interaction between the roles leads to extensive self-organisation. For this to work, clear and harmonised roles are required. Cooperation within the strategy team and the roles are described in detail in Chapter 6. Below is a brief overview of the roles:

Enabler and Strategy Owner
The Enabler provides the boundaries for agile strategizing. They set the strategic options and, together with the Strategy Owner, defines the Strategy Orientation parameters (North Star with mission, vision/guiding principles and values). The Enabler is responsible for the company's success and is located at the top management level of the organisation. The Strategy Owner is responsible for a consistent and effective strategy. To this end, they develop strategy variants, commission a review of the underlying assumptions and select the most effective strategy with its individual elements.

Enablers and Strategy Owners focus on the impact of the strategic option and manage the use of resources in relation to the impact to be achieved. The primary objective of this impact management is to ensure the long-term sustainability of the organisation and the achievement of the overall corporate goals.

Hypothesis Owner and Strategy Exploration Team
The Hypothesis Owner and Strategy Exploration Team concretise the strategy elements. They define the hypotheses and tests based on the chosen strategic option, execute and analyse them. Thanks to the empirical approach, they generate a continuous increase in knowledge with each iteration. As the level of maturity of the Strategy Development increases, they confirm or refute the prospects of success of a strategic option.

The Hypothesis Owner and the Strategy Exploration Team are responsible for the quality and significance of the information obtained.

Strategility Master
The Strategility Master supports and advises everyone involved in the Strategility process. They define the Strategility process together with the participants and facilitate and adapt the strategy process. They take responsibility for the process and ensure transparency and feedback of the insights gained.

Collaboration: Participate ≠ Co-Decide
Agile strategizing relies on targeted collaboration and the involvement of many in order to achieve greater strength in implementation, to have a greater diversity of ideas or to be able to act in a network with other organisations. This open type of strategizing can take place within an organisation, a company or across organisational boundaries. The term «open strategy« is a collection of complementary approaches and techniques.

The most common stumbling block when involving some or many people in strategizing is having exaggerated expectations. Participants often overestimate their influence and expect to have a say in everything. Strategility's simple role model helps to clearly define tasks and manage expectations. Simplified: Contributors are typically part of the Strategy Exploration Team and have the task of developing the basis for decision-making.

In principle, the role of the Enabler can freely shape the decision-making process. The important thing is that the scope for decision-making remains within the range of feasible decisions. An organisation with a legal mandate must stay within its mandate. A sports club should stay within the field of physical activities. A company can decide to produce mobile phones instead of rubber boots if the owners are prepared to accept this change of direction.

Collaboration: Clear Intent and Clear Expectations
It is the responsibility of the Enabler and the Strategy Owner to formulate clear intentions for involving many: What problem in strategizing do they want to solve by involving some or many? It is then the task of the Strategy Master to create the meetings, interactions, etc. to achieve this.

If required, open strategizing can be used throughout the entire Strategility flow. More important, however, is the question of the impact to be achieved through the design and inclusion of appropriate interactions or workshop formats. Inclusion also includes a visible impact of the contribution. Transparent communication about the use of the results is important here. In particular, there should be clarity about any suggestions that are not followed up. A clearly communicated subjective individual decision by a manager can be just as transparent as the disclosure of analyses and considerations.

The long-term nature of strategizing makes it difficult for contributors to recognise their contribution. Therefore, make sure that parts of the contributions in the contributors' sphere of influence are implemented immediately and directly, e.g. implementing a value from the new mission statement in everyday life, or seeking customer proximity in concrete terms, eliminating a low value-added activity, acquiring digital skills, organising meetings, and gatherings differently, etc.

Expect contradictory behaviour with regard to participation. There is often a strong demand for information and participation. At the same time, compulsory events are rejected. Voluntary participation combined with low-threshold formats have therefore proved successful.

Low threshold means being able to contribute with little effort. Generally accessible spaces such as cafeterias, entrance halls, and corridors can be used for low-threshold contributions – either as «workspaces« for workshops or by means of posters with simple interaction options. To organise information events as co-creation events with mini-interactions, mumble groups or online surveys have proved effective. If the information event is voluntary, it is possible to participate remotely and with a time delay. This promotes both voluntariness and the contribution threshold.

Practical Tip

The Right Question

Most organisations operate within fixed boundaries. Suggestions, feedback, and impulses outside this boundaries cannot be followed up. The narrower the boundaries, the more focused the questions asked should be. For example, instead of asking for feedback on a new vision, you might ask three specific questions about the vision. It is also a good idea to ask employees to elaborate on a rough sketch of the vision or to present alternatives for discussion.

Strategic thinking is difficult for many people. Enabling participation at a practical, operational level can deliver richer results. For example, when developing a new vision, three questions can be helpful:

1. How does your current work contribute to the new vision?
2. How/where can the vision help us to make better decisions in our daily work?
3. What will it take for you to identify with the vision?

The questions encourage a serious examination of the vision and leave room for manoeuvre. You can use the feedback to correct the vision, strengthen its implementation or initiate change measures.

2.7 Steering

2.7.1 Review & Planning Event

At the heart of Strategility steering is the «Review & Planning« event (see Figure 2.7), whose main instrument is the Strategility Board. The «Review & Planning« event gives strategizing a regular rhythm. This event marks the end of a Strategility iteration (see Chapter 2.7.2) and heralds the start of the next iteration. The agenda consists of the following main points, which are described in detail in the following sub-chapter:

1. **Review:** Gaining insights from the results and progress of completed strategizing
2. **Process improvement** (retrospective): Working methods, collaboration, and procedures
3. **Decision:**
   - Accept or reject requests
   - Accept or reject results
4. **Planning:** Plan strategizing for the subsequent iteration
5. **Prioritisation:** prioritise strategizing
6. **Release:** Release the plan for the subsequent iteration

The Strategility Board (see Figure 2.11) is used for steering.

Event
Strategility Review & Planning
Goals:

- Create a shared view of the current state of insights, progress, milestones, and outputs
- Lessons learned and decisions made
- Prioritise and plan the next steps

Participants:

- Facilitation: Strategility Master
- Decision: Enabler
- Contributors: Strategy Owner, Hypothesis Owner, Strategy Exploration Team
- Guests as required: implementation owners, experts

Work steps:

- Brief presentation: strategic challenge and current facts
- Presentation: Results of the completed Strategility iteration
- Workshop interaction: Insights: Assessing assumptions and serious consequences – «What have we learnt about the assumptions and impact of our strategic decisions?«
- Workshop interaction: Process improvement in strategizing – «What do we need to change in the strategic positioning/strategy/implementation project?«

- Decision: Objectives and priorities for the subsequent iteration – «Which tasks do we include in the Strategility Board?«
- Workshop interaction: Rough planning of the subsequent iteration – «Which tasks should be tackled?«
- Decision: Prioritisation of the upcoming strategizing – «Which work now provides the greatest added value/gain in insight?«

Results and outputs:

- Next iteration planned
- Dissemination of results: Strategy Orientation, strategic option, strategy, implementation mandates, ...
- Updated Strategility Board
- Shared understanding of lessons learned
- Improved cooperation, procedures, processes

Review & Planning includes strategizing for all active strategy processes: Strategy Orientation, Strategy Development, and Strategy Implementation. Review & Planning takes place at regular intervals. The quarterly cycle is often chosen. More detailed considerations on the choice of cycle can be found in the following explanations on the iterations.

You can save time by working on Strategy Orientation, Strategy Development, and Strategy Implementation simultaneously. This can be extremely attractive when the urgency is high. Therefore, the activities in agile strategizing are risk-focused: systematically eliminating uncertainty and gaining targeted insights in order to gain decision-making certainty. In addition to this, the negative effects of unfavourable decisions are reduced by short iterations.

To ensure that the agile principles (see Chapter 7.1) have the greatest effect in the face of high uncertainty and a need for speed, it is advisable to conduct the review and planning with all those involved in strategizing in the form of big room planning.

Practical Tip

Big Room Planning

Big room planning is an interactive large group plan developing event. «Big room« means that the plan is developed together and visualised on a large wall using sticky notes.

In the spirit of an all-hands-on-deck event, all participants in the «big room« work together on a plan. In agile software development, big room planning sessions are held for quarterly plans with up to 250 people over two days. We have also had good experience with big room planning in core teams or with delegates rather than the whole workforce.

Planning a strategy iteration in a team of up to 10 people can be expected to take one hour with experienced facilitation. Planning a two-week sprint in a Strategy Exploration Team will take two hours.

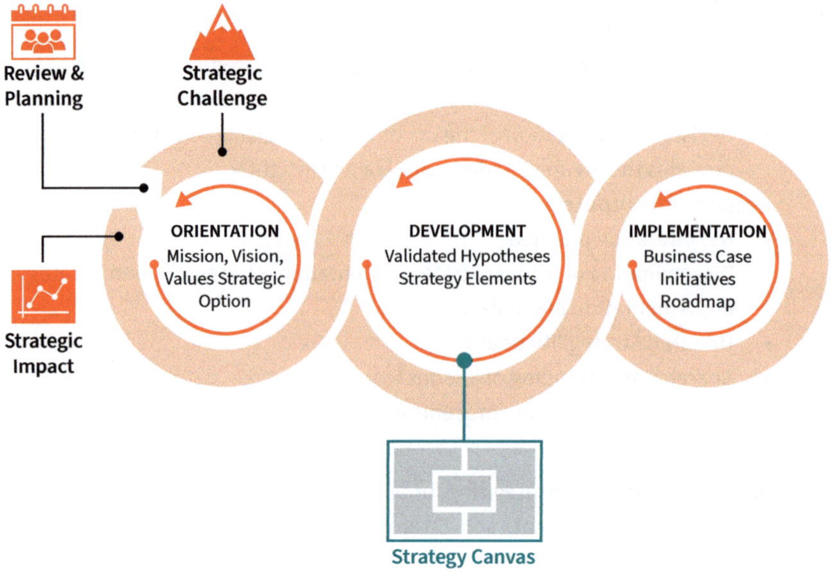

Fig. 2.7: Strategility event «Review & Planning«

Example

Parallel Strategizing: Renting in the Department Store

In our example of the department store testing «renting instead of buying« as a new strategic option, strategizing was structured roughly as follows after eight months:

Strategic challenge:

Reversing the decline in bricks-and-mortar sales and achieving growth

Steering with Review & Planning:

Every two months, the Hypothesis Owner holds a regular meeting with the management (Enabler and Strategy Owner), involving members of the Exploration Team and other experts on a case-by-case basis.

The Strategy Orientation sub-process:

Validation of the «renting instead of buying« strategic options in product development

Strategy Development as a sub-process in the Exploration Team:

Enabling growth and determining sustainability:

- Exploration Team of five people
- One week Scrum[4] sprints
- Definition, implementation, and evaluation of hypotheses and tests in the Exploration Team's backlog:

4 www.scrum.org: Agile development method for complex problems.

  - What is an effective rental offer? Expanding the rental offering
  - What are effective core competencies and value-added steps for the rental offering: make-or-buy decision for appropriate reverse logistics for goods returns, disposal, and recycling?
  - What are effective methods of business development? Collaboration with suppliers for selected product groups

The Strategy Implementation sub-process:
Operate rental in a partial implementation as a pilot project and analyse the impact:

- Pilot rental operation led by members of the Exploration Team
- Working with a Kanban Board and fortnightly planning iterations
- Tasks:
  - Supporting field and customer service staff
  - Handling returns
  - Operation of the experimental online rental shop
  - Create a business plan and project canvas for full implementation

2.7.2 Strategility Iteration

Agile strategizing is divided into fixed time periods, called iterations. The fixed iterations form the backbone of strategizing. They are driven by the Review & Planning event (see Chapter 2.7.1): Planning heralds the start of an iteration and the review concludes it. In the Strategility iterations, strategic insights are reviewed for their impact on the three process parts «Strategy Orientation«, «Strategy Development«, and «Strategy Implementation«. The regular cycle creates fixed structures for Strategy Development and Implementation. This leads to predictability and therefore planning certainty in strategizing.

A Strategility iteration lasts between two and six months. In the case of high uncertainty, high complexity and rapid change, a shorter cycle should be chosen. A short cycle is also recommended when there is great urgency. The cycle, i. e. the length of the iteration, determines the agility and the maximum loss of time and resources in case of unfavourable decisions. The short cycle is balanced by the inertia of the system: Decision making and implementation time (see Figure 2.8).

Fig. 2.8: Tension field short vs. long iteration time

2

Practical Tip

Length of the Iteration

Most organisations strive for a quarterly rhythm of iterations and fail either because the pace is too fast or because of the changing intensity of work due to holiday periods and year-end work in the calendar year.
Three trimesters of Review & Planning in winter, spring, and autumn, with different lengths but with a more constant capacity, is often more practical.

Big Leap

In the face of major uncertainties or in a complex environment, the effort required to achieve planning certainty for the «big leap« – i. e. a comprehensively conceived change in strategy – is very high. Due to the dynamics of the environment, industries, and company events, the basis for decision-making and thus the plans become outdated faster than the big picture can be realised. As a result, the risk associated with the tried-and-tested, plan-driven approach is high, due to the constantly high probability of making the wrong decisions and wasting a lot of time and resources (see Figure 2.9).

Example

«Big Leap« Logistics Centre: Rental in the Department Store

Efficient logistics are important for a department store. Rental places new demands on infrastructure and processes: Reverse logistics is becoming increasingly important. This includes the return, refurbishment, and, if necessary, repair or sale of goods via outlets or other channels.
Setting up an efficient reverse logistics system is time consuming and capital intensive. Without a validated strategic option of «product development for returnable assortments«, setting up reverse logistics as a «big shot in the dark« can lead to high losses. Outsourcing reverse logistics to a service provider would be a low-risk, scalable alternative. The disadvantages of outsourcing in the event of success must be weighed up against the damage caused by in-house investment in the event of failure (see also the comments on «Mechanical Turk« in Chapter 8.6).

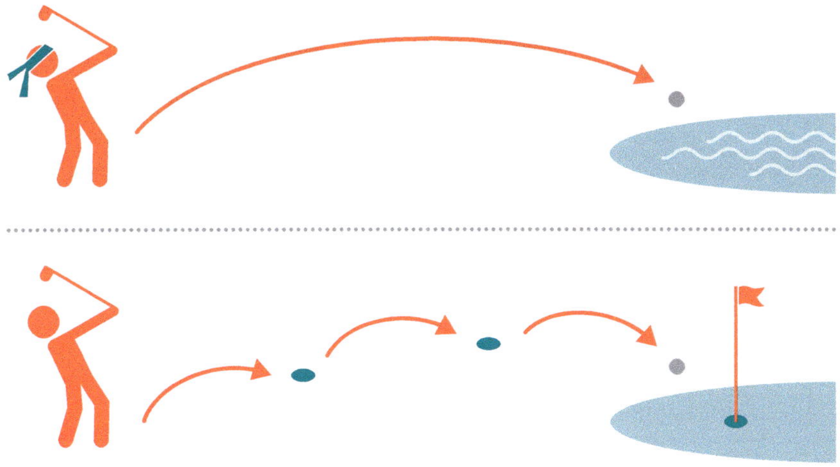

Fig. 2.9: Big risk in a big leap

Practical Tip

Stumbling Block Comfort Zone

Because established structures and processes are designed for predictability and longevity, organisations are unable to cope with short iterations. The necessary adjustments to working methods and ways of thinking are met with resistance. To avoid the pressure to change, longer iterations are chosen, which are more in line with established ways of working with large casts and few decisions.

Avoid that! Shift your attention away from the results of your work and towards the lessons learned. Keep these learning iterations short. In this way, you promote a learning culture and an agile mindset in strategizing.

Take the example of «renting instead of buying«: short-term experiences from customer interactions, such as experimenting in a shop or pop-up store, are preferable to lengthy concepts for logistics and rental ranges.

2.7.3 Orientation in the Steering System

Guidance and navigation need orientation. Where should the journey take us? In strategizing, the strategic challenge provides the necessary orientation (see Figure 2.10):

- **Prospects:** Concise statement of current focus
- **Target:** Target values of observables or impact indicators. Observables are often qualitative.
- **Actual:** Actual values of observables or impact indicators

	Initial Position	Driver	Consequences	Obstacles	Success
Prospects					
Target					
Actual					

Fig. 2.10: Orientation in the steering of strategizing

2.7.4 Steering with the Strategility Board

In the Strategility Board (see Figure 2.11) you record all processes in the flow of strategizing in Strategy Orientation, Strategy Development, and Strategy Implementation. A status is recorded for each process. Make the Strategility Board accessible to all stakeholders. This transparency of all processes promotes shared alignment and gives autonomy to individuals and teams. Autonomy as an agile principle promotes flexibility and speed.

The states of the processes on the Strategility Board are:

- **Backlog:** Ordered list of pending work, with the first element being the most important and bringing the greatest added value in the shortest time (see also the explanations on prioritisation in Chapter 2.1)
- **In planning:** Necessary planning work, e.g. to estimate the value added or the duration and cost of work
- **In progress:** Ongoing strategizing across the three strategy processes of orientation, development and implementation. Tasks are finalised through an approval process. The approved results – for example a formulated strategy – is then considered binding.
- **Under observation:** Ongoing impact measurement of results. For example, sales may be reorganised as part of market penetration. After this implementation, it is advisable to monitor the success and impact of the sales and to use the insights to correct the strategizing process.
- **Completed:** Completed tasks that require no further monitoring

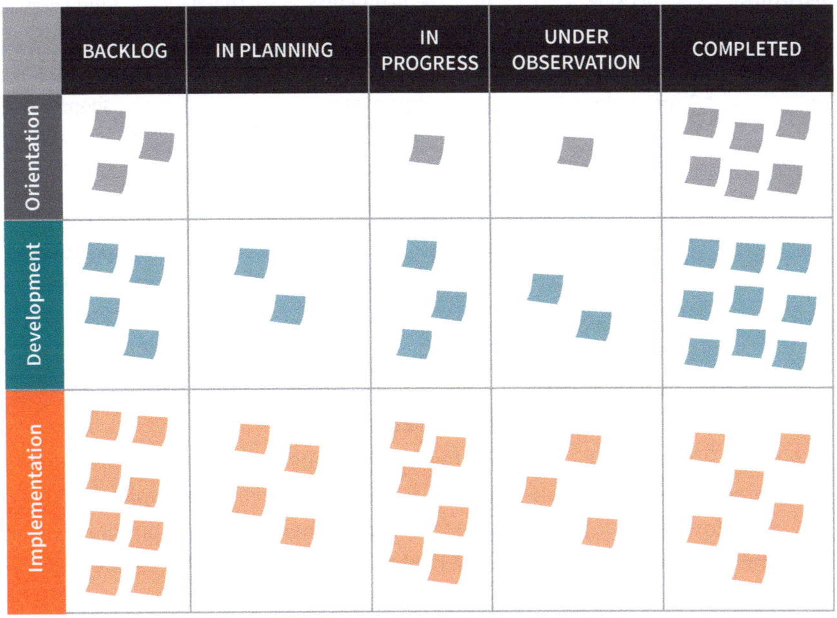

Fig. 2.11: Strategility Board

Table 2.3 illustrates a simplified Strategility Board after about eight months of strategizing using the example of «renting instead of buying« in the retail sector. In the table, the number of tasks has been reduced to about one fifth.

Department store	Backlog	In planning	In progress	Under observation	Completed
Orientation	1. Digitisation of space 2. AI shopping app 3. Shopping experience		Retail in the sharing economy	The strategic challenge of bricks-and-mortar retail	
Development	1. Make-or-buy decision for reverse logistics 2. Expanding the rental offer 3. Collaboration with suppliers		Formulate strategy	Collect and review impact indicators	1. Field test customer interest 2. Pop-up store: Finding your target audience

Department store	Backlog	In planning	In progress	Under observation	Completed
Implementation	1. Operationalise production 2. Implement secondary use 3. Marketing: Positioning messages		Create key figures and reporting		1. Fake online shop: Operational experience 2. Fake process: provoking growth

Table 2.3: Example Strategility Board «Renting instead of buying«

Practical Tip

States in the Strategility Board

The states on the Strategility Board are used for control and transparency in the process. Use your own states and labels. States typically form a workflow from left to right. More states mean finer control options or insights into the processes. The minimum configuration comprises three states: backlog, in progress, completed.
More than three states are common and recommended for managing implementation projects.

In the Strategility Board, you manage different «swim lanes« for the processes of Strategy Orientation, Strategy Development, and Strategy Implementation. This gives the responsible individuals and teams sovereignty over their processes and provides room for different speeds:

- **Swim lane orientation**
  The Enabler and the Strategy Owner are responsible for the work.
  The work on Strategy Orientation is typically completed in one or two iterations. They can then remain in the «under observation« status for a long time – over many iterations.
- **Swim lane development**
  Strategy Owners, Hypothesis Owners, and the Strategility Exploration Team are responsible for these tasks.
  Strategy Development tasks are completed within the iteration.
- **Swim lane implementation**
  Overall supervision of the Strategy Implementation typically lies with the Enabler. The Enabler may delegate the supervision of individual initiatives.
  Responsibilities for concrete implementation vary depending on the type and form of implementation (see also Chapter 6).
  When tactical measures are implemented in the line organisation, department or division heads are responsible for the implementation initiatives.

In the case of implementation via project organisation, responsibility for implementation lies with portfolio, programme, and project managers.
Depending on the type and working method – agile or plan-driven – Strategy Implementation initiatives can remain in one or a few iterations or in «work in progress« status for a long time.
For example, the construction of a new logistics centre required for market development may take years. This implementation initiative would have the status «in progress« throughout the entire construction period.

> **Practical Tip**
> **Kanban for the Strategility Board**
> The Kanban method (Anderson 2011) is popular and proven for agile processes. We recommend using it together with a Kanban Board. The Strategility Master acts as the process owner in Kanban. Depending on the chosen Kanban method, they are called Flow Master, Flow Manager or Delivery Manager, Lean or Lean-Agile Master.
> We also recommend using the Kanban metrics of the number of processes – work in progress (WIP) – and the lead time of the processes to increase the effectiveness and efficiency in strategizing.

The Strategility Master is responsible for ensuring that the board is up-to-date and consistent. They are also responsible for ensuring that metrics, indicators, and observations are up to date. Agile strategizing is based on self-organised teams and individuals acting on their own responsibility. The Strategility Master should therefore not be the one to collate and update the board or the metrics themselves. Experience shows that a «master of ceremonies« helps to remind and empower colleagues to maintain the boards and keep them up to date.

2.7.5 Linking Strategy Development

Steering Strategility at regular intervals, typically every three months, via the «Review & Planning« event, the iteration, and the Strategility Board bring together the work in Orientation, Development, and Implementation.

For the actual strategizing in Agile Strategy Development in the Exploration Team, the team works at its own pace with its own events and its own boards.

For Agile Strategy Development in the Exploration Team, we recommend a two-week cycle, the strategy sprint. This starts with a sprint planning event and concludes with a sprint review event. Between two sprints there is a combined sprint review and planning event. Iteration planning is therefore followed by multiple sprints with multiple planning and review events. A hypothesis canvas (see Chapter 4.3) can be used to manage the work in the strategy sprint. If required, the hypothesis canvas can incorporate further work by the Exploration Team. The consolidated outputs and insights from the sprint reviews flow into the overarching strategy review. Conversely, the releases from Strategility Planning form the basis for sprint planning (see Figure 2.12).

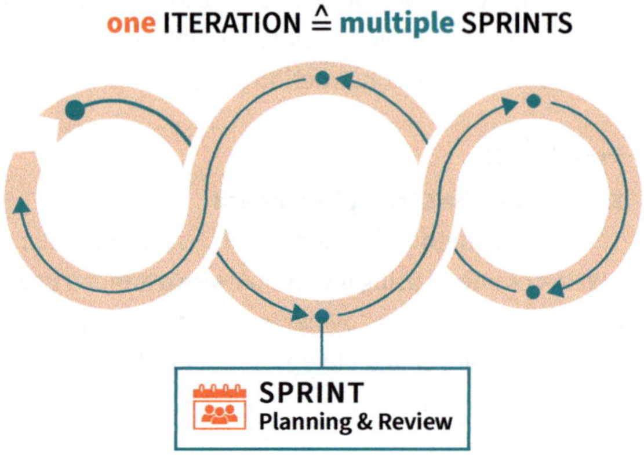

Fig. 2.12: Linking Strategility iteration and strategy sprint

The task of the Strategility Master is to moderate the events and guide the process.

CHAPTER 3

Agility for Strategic Orientation

3 Agility for Strategy Orientation

Strategy Orientation is about formulating the North Star of your organisation, which will guide you on the long run through turbulence and short-term uncertainties. You should identify the strategic challenge for your organisation and formulate a strategic option to address it. Agile Strategy Orientation ensures that you have regular insight into the impact of the strategy and the implementation initiatives, allowing you to reflect on the suitability and future orientation of your organisation. If necessary, the strategic option can be corrected before resources are invested in its implementation. The output is a strategic option that provides orientation.

If you develop your Strategy Orientation according to the principles of Strategility (see Chapter 1), you will accomplish the following in this sub-process (see Figure 3.1):

- Alignment of the owner's strategy or performance mandate with the Strategy Orientation parameters: Mission, values, vision, and strategic option (alignment)
- Feedback of insights from Strategy Development and Implementation to review the strategic option (evidence)
- A process that repeatedly triggers knowledge acquisition and reflection on the strategic option (iteration)
- Collaboration between experts and decision-makers in order to jointly develop a basis for decision-making (collaboration)

In the process of Agile Strategy Orientation, you derive the strategic challenge for your organisation or administration from the analysis of your initial situation. You develop strategic options to respond effectively to this challenge. You evaluate these and regularly reflect on their effectiveness based on the lessons learned from Strategy Development and Implementation.

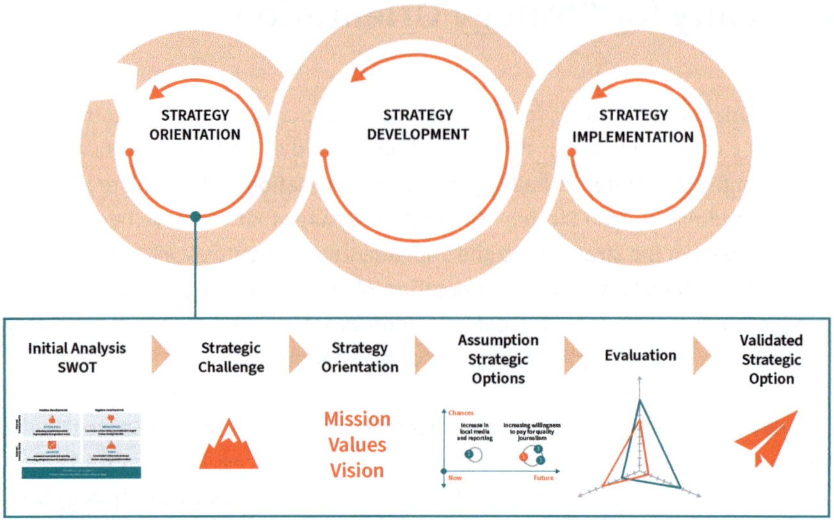

Fig. 3.1: Sub-process of Strategy Orientation

3.1 Identify your Strategic Challenge

To identify which strategic challenge requires the attention and resources of your organisation and administration, you can use the established methods of strategic management for the initial analysis (see Johnson et al. 2018, p. 60 ff.). They support the examination of changes that are taking place inside and outside your organisation and that have a significant influence on your organisational development. The PESTEL analysis, which analyses political, economic and social developments, among others, has become established as an overarching environmental analysis. Porter's Five Forces Analysis (Porter 2008) is used for an industry analysis and a value creation analysis of the processes for an internal analysis (Rüegg-Stürm/Grand 2020). This is not an exhaustive list, but rather an indication of common methods of initial analysis that remain relevant in an Agile Strategy Orientation.

The insights from these initial analyses are usually summarised in the so-called SWOT analysis. A template for this can be found in the *Digital Extras*. Internal strengths and weaknesses as well as external opportunities and threats are derived from the above analysis.

At this point, we recommend that you draw a conscious conclusion: What is the real strategic challenge for your organisation that you need to address in the next two to four years? To do this, Enablers and know-how providers should work together to summarise the orientation aspects for management (see Chapter 2.7):

- What do we consider to be important factors in our **starting position**?
- What **changes** in these initial factors do we consider relevant for future development?

- What **prospects** do we have if we do not change our current direction and strategy?
- What **uncertainties** do we have about reorganisation?
- How would we recognise the **success** of a future direction?

To summarise, a clear statement can be derived from this: What is our strategic challenge? This will help you to focus on the impact of your strategic activities and to keep them under review: Do the strategic decisions help to answer the strategic challenge?

Example

The Strategic Challenge of an Online Newspaper

An online newspaper was founded with the aim of providing investigative journalism that is independent of sponsorship and advertising. The initial analysis revealed the following factors as relevant to success:

- Public interest and willingness to pay for quality journalism
- Credible ethical values within the newspaper
- The newspaper's own strategic positioning
- The number and financial strength of the newspaper's subscribers

There was a chance that the newspaper would find subscribers willing to pay; at the same time, details of their interest and scope were uncertain. However, if sufficient new subscribers were not attracted, it would be impossible to achieve the growth and financial security necessary for medium-term survival. As a result, the newspaper would either have to abandon its core value of advertising freedom or go out of business.

The strategic challenge was derived from these aspects: «We must achieve lucrative growth in the coming years without compromising our ethical values.«

Fig. 3.2: SWOT and the strategic challenge of an online newspaper

The strategic challenge identified may be so comprehensive that it changes your entire strategic direction. Or it may only affect certain aspects of the strategic direction. Below we explain which Strategy Orientation parameters may be affected in each case.

3.2 Mission Statement and Strategic Options – Contents of Your Strategy Orientation

Strategy Orientation – Orientation Parameters

The Strategy Orientation describes the longer-term development horizons of your organisation and administration:

- For example, are you focused on long-term survival – because you are a family business – or on short-term profitability – because, as a start-up, you are aiming for successful market positioning and an early sale to an established company?
- Will your authority focus on growth over the coming years – for example, because the economic sector in your region needs to be boosted – or on efficiency – because the financial budget needs to be consolidated?

Such strategic directions of an organisation are usually described by the four Strategy Orientation parameters of mission, values, vision, and strategic options. They provide orientation for focusing activities and resources. They must, of course, be aligned with the owner's strategy or the legal mandate and describe the scope for action within these guidelines[5]. The strategic options are the link to Strategy Development.

- The **mission** describes the orientation of an organisation as an organ of this society: What is our contribution to this society, what is our mission, why are we here?
- The **values** describe the basic attitude with which all members of an organisation and administration should act. They provide normative guidance in uncertain decision-making situations, for example: When in doubt, do we act for stability, visibility, and order, or for creativity, subjectivity, and free development (see Chapter 7)?
- The **vision** describes an organisation's position in the future: What developments do we see in politics, the economy, and society, and how should we position ourselves within them?
- **Strategic options** describe the direction of indirect organisational development: Do we want to enter new service areas and markets or optimise our existing offerings and markets? We present typical strategic options below.

Mission Statement, North Star, and Purpose

Mission, values, and vision are also often referred to as the mission statement or North Star. In recent years, the term «purpose« has also been used increasingly. Purpose is understood as a meaningful objective that goes beyond profit maximisation. Many organisations continue to work

5 For examples and the development of these benchmarks in public administration, see also: https://www.olev.de/l/leitbild.htm.

with the guiding principles of mission and vision, which is why we also use these common terms below.

The Strategy Orientation parameters for an online newspaper are shown clockwise in Figure 3.3. You can find an example of Strategy Orientation parameters for a youth and careers service and a template for outlining your own orientation parameters in the *Digital Extras*.

Example

Strategy Orientation of an Online Newspaper

An online newspaper positions itself in society and the economy with the mission of being an independent antithesis to large media groups. Accordingly, it positions itself in the canon of values with transparency, diversity of opinion, independence, and freedom from advertising. Its vision is to be a recognised information authority on controversial topics in five years' time. Its strategic option is therefore to increase its penetration of its current market. This will enable it to meet the strategic challenge of achieving growth without losing its values of independence and freedom from advertising.

Fig. 3.3: Strategy Orientation of an online newspaper

Addressing Strategic Challenges in the Strategy Orientation

Explicit statements on these four Strategy Orientation parameters make it possible to discuss the strategic challenges and appropriate orientation emphases with owners, legal clients, and members of the organisation at the «right level«, for example:

- **Mission:** Do we want to remain an online newspaper in any case, or should we also distribute printed information in the case of strategic challenges – for example, in the case of acute grievances that need to be widely publicised, or in the case of a relevant readership that does not read online, etc.?

- **Values:** How much is freedom of advertising worth to us? Would we sacrifice it for the survival of the newspaper in the face of the strategic challenge of a ruinous level of debt?
- **Vision:** How many resources can we raise as a pre-investment for the targeted five-year growth? Would we change the vision if faced with a strategic challenge such as employee turnover or lack of investment funds?
- **Strategic option:** Should all our activities be focused only on «hot topics« or would we perhaps also direct resources towards promoting everyday issues in the strategic challenge of a «social depression«?

The mission, values, and vision are usually developed jointly by the owners, the board of directors, the management, and often also the staff. Ideally, they are plausibly derived from an initial analysis. A more detailed procedure for evaluating the mission, values, vision, and strategic options against the background of the strategic challenge is described in Chapter 3.4.

While mission and vision tend to describe fundamental, longer-term orientations, strategic options need to be reviewed regularly. The strategic option is therefore the planned start of Agile Strategy Development, the strategy its output. We describe the strategic options in more detail below.

3.3 Strategic Options – Responses to Strategic Challenges

Strategic options set out the long-term vision of an organisation without providing detailed specifications. They help to define the broad direction of development in response to current strategic challenges over the next two to four years, for example: «The competition in our market is reducing our market share, so we want to penetrate our market more strongly.« Or: «Politicians have other service priorities for the next few years, so we want to optimise our range of services.« This provides a useful level of abstraction for strategy discussions. In practice, however, we often find that these strategic options are not explicitly formulated, both in organisations and in administrations. In the following, we argue strongly in support of these often not explicitly formulated orientation parameters and explain different variants of strategic options.

Strategic options are commonly divided into growth and optimisation options. This rough categorisation can be illustrated using an extended Ansoff matrix[6]. The logic of this structuring of strategic options can also be applied to the public sector. We adapt the corresponding terminology below. Figure 3.4 shows possible strategic options for the online newspaper; in the *Digital Extras* you will find a template for your own strategic options. Examples for public administration are given below.

6 The matrix is named after its developer Harry Igor Ansoff. It was originally developed to visualise and estimate growth potential (Ansoff 1965). We have expanded it to include the strategic option of withdrawal from markets.

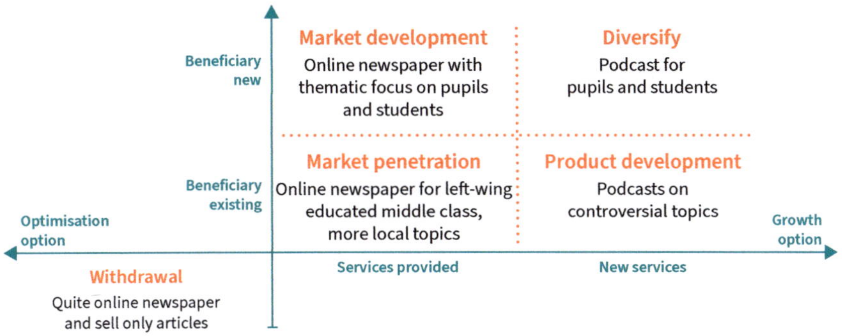

Fig. 3.4: Strategic options of the online newspaper

The strategic option you choose for your organisation should depend on the attractiveness of the market and your internal capabilities. In the following, we outline the strategic option and framework conditions:

- **Withdrawal – in administration: privatisation**
  In the administration you withdraw from service areas such as childcare or municipal housing management. In the private sector, you are withdrawing from customer or product segments, possibly an entire industry.
  The strategic framework conditions that suggest such a decision are shrinking markets. In the public sector, service areas may be provided by private companies in a more needs-based, cost-effective, and socially responsible manner.
  Don't necessarily see the strategic option of withdrawal as a failure or the «beginning of the end«. Not only can it allow you to withdraw from unattractive service areas, it can also help you to focus your resources and thus become more differentiated and successful. For example, one project management company withdrew from the unprofitable product range of «sales of project management tools«. The company has been successful for many years by focusing on training and consultancy on project management tools.
- **Market penetration – in administration: service intensification**
  You remain in your traditional service and customer environment. The aim may be to maintain or increase your own performance or market share or to occupy an important niche.
  Strategic framework conditions that suggest this strategic option are present when growth is necessary, but internal resources do not allow the development of new additional services, markets or products. Market penetration or service intensification can also be a response to competitors that you want to and can displace, or to customers with whom you want to position yourself in a particular way.
  For example, a youth and career counselling service offered targeted career counselling to single parents. In doing so, it has intensified its existing career counselling services for a selected customer group in order to expand its preventive positioning.

- **Product development – in administration: service development**
  You are entering a new service or product area. Services that you have not previously offered are intended to create new benefits for existing customers and new added value for your organisation. Digitalisation has been the driving force behind product development in recent years.
  The strategic challenge that this strategic option addresses is the desired growth that cannot be achieved with the current products. The framework conditions that suggest this strategic option are attractive customer groups that should continue to be bound to the organisation and the opportunity to create added value for customers and the organisation itself by expanding the range of services. This can be done by adapting to current service standards or by offering attractive additional services. Of course, this strategic option requires innovation and service provision capabilities.
- **Market development – in administration: additional beneficiaries**
  You are entering into a new customer environment. You are offering existing, successful products and services in new geographical regions or to previously untapped customer groups.
  The strategic challenges here are usually mature markets where further growth is no longer possible. Conditions that suggest this strategic option are strong products and services as well as attractive, untapped customer groups. In addition, communication and marketing capabilities are needed to successfully address these new customer groups.
- **Diversification**
  This strategic option combines new service offerings and new customer groups and therefore represents the most complex and risky direction of development for an organisation. Diversification can have varying degrees of connection to an organisation's existing business:
  a) It can take place within the traditional value chain – through the integration of supplier or customer services. For example, the online newspaper could set up a database for the exchange of local contributions.
  b) Diversification can take place within its own value chain, by expanding its offerings and customer groups within the same sector. For example, the online newspaper could offer news blogs for clubs or companies.
  c) Diversification can be unrelated to the company's existing business. For example, the online newspaper could set up a news café.

  Strategic challenges that suggest this strategic option of diversification are changing mandates and expectations of owners and the need for a broader distribution of risk. Helpful conditions include sufficient competencies in innovation, service delivery and marketing, and strong management to create and steer added value between the emerging business areas.

In the *Digital Extras*, you will find an overview of which of your strategic challenges and conditions correspond to which strategic options for organisations and the administration. For a conscientious analysis and definition of your strategic option, we recommend that you carry out the usual steps of an initial strategic analysis of external and internal company conditions (see Chapter 3.1 and Johnson et al. 2018, p. 60 ff.).

The strategic option is usually developed and approved by the management – i. e. the Enablers. Ideally, it is plausibly derived from an initial analysis. A more detailed procedure for evaluating strategic options against the background of the strategic challenge is described in Chapter 3.4.

The decision on the strategic options outlined above will be further refined in the Strategy Development process (Chapter 4).

3.4 Details of the Strategy Orientation Process

In the following, we present two steps and methods for developing the Strategy Orientation, which can help you to derive and evaluate the orientation parameters from the initial analysis.

Assumptions for an Effective Strategy Orientation

The agile development of your organisation's orientation can be triggered by two events:

- Changes in the strategic starting position that lead to a strategic challenge (see Chapter 3.1)
- Strategy Development insights that indicate that your Strategy Orientation is not effective in addressing the strategic challenge (see Chapter 4)

Based on these impulses, you develop a mission statement and a strategic option for the Strategy Orientation of your organisation in order to respond effectively to the strategic challenge.

To do this, you can use the SWOT^3 method (Thode/Wistuba 2019) to map the external and internal influencing factors on a coordinate system (Figure 3.5). The external influencing factors are positioned horizontally on a time axis as circles according to their probable time of occurrence. The size of each circle symbolises its importance to your organisation, and their vertical position represents their assessment as either an opportunity or a risk. The relevant internal drivers are assigned to each external driver. This differentiated presentation provides a simple overview of strategic fields of action, the significance, and timeframe of which can be adjusted as new information becomes available. A template for the SWOT^3 can be found in the *Digital Extras*.

The example of the online newspaper (Figure 3.5) shows that the strategic risk of market consolidation is countered by the internal weakness of an unclear strategic direction – an aspect that urgently needs to be mitigated by a vision and strategic option that provide orientation. If the online newspaper cannot position itself clearly in a market by a few strong competitors, it will barely be recognised and could lose potential subscribers to the major competitor.

On the other hand, the external trend of market consolidation, following a successful strategic positioning of the online newspaper, could probably even be classified as an opportunity in the next SWOT³ version, because it would stand out in a homogeneous media landscape.

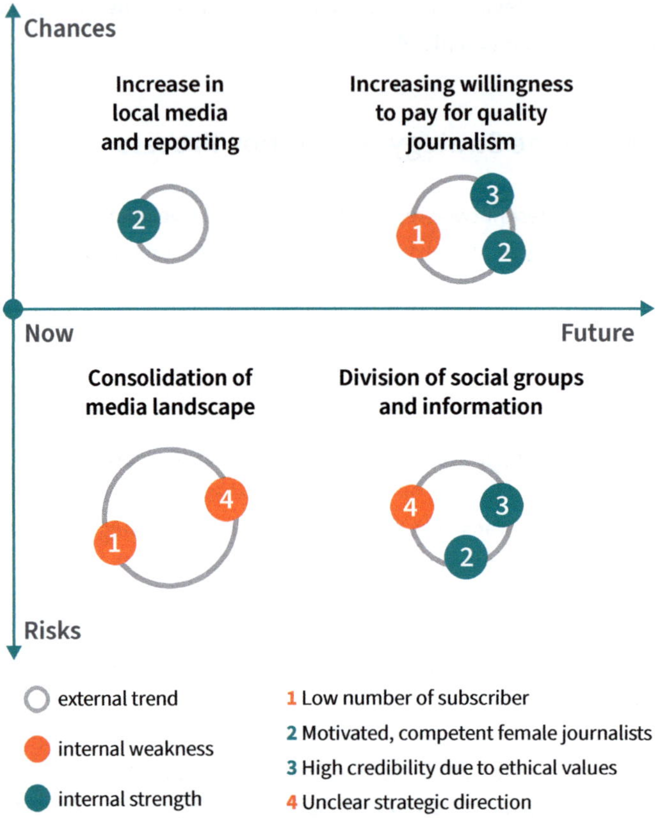

Fig. 3.5: SWOT³ for the online newspaper

Your derived mission, vision, and strategic option are the presumably sensible responses to assumed developments. They are based on assumptions about the future in society, the economy, nature, technology, and politics as well as on assumptions about the positive impact of the defined organisational direction on the strategic challenge to be solved.

Make these assumptions consciously. For example, you can use scenarios in which important influencing factors are explicitly assumed in various forms and the strategic direction of the organisation is formulated for each scenario.[7]

For an evidence-based definition of your strategic direction, you should formulate parameters to be tracked, called early warning indicators, that will give you an indication of whether the development of the influencing factors will actually materialise.

You should also define indicators that provide you with information on the positive impact of your Strategy Orientation.

> **Example**
> **Early Warning and Impact Indicators of an Online Newspaper**
> If the online newspaper were to assume a scenario in which the media landscape is limited to three dominant media houses in the next five years, an early warning indicator could be whether currently independent newspapers are being bought up by dominant media houses.
> In the example of the online newspaper, an impact indicator could be the increase in subscribers after providing increased coverage on controversial local issues.

The testing of these assumptions against the early warning and impact indicators should be entered in the «in progress« column of the hypothesis board and in the «under observation« column of the Strategility Board. In the Review & Planning events, the development of the indicators should be analysed at least once every quarter (see Chapter 2.4). The Strategility Master orchestrates the compilation and analysis of the data, and the Enabler decides on their impact on the Strategy Orientation and how to proceed in terms of strategic option and Strategy Development.

Evaluation of Potential Strategic Options

In order for the strategic options to be approved by the Enabler as a benchmark for Strategy Development, they are again evaluated by the Enabler and strategic experts from the organisation – such as the potential Strategy Owner. Suitability, financial acceptability, and feasibility have been established as evaluation criteria in strategic management (see Johnson et al. 2018). These criteria can – where feasible and meaningful – be assessed analytically; in the case of financial acceptability, for example, with break-even or discounted cash flow analyses (cf. Johnson et al. 2018). In our experience, however, many reorientations are difficult to analyse, which is why we recommend an argumentative discussion based on the facts provided by important key experts.

7 For tips on scenario development, see https://www.cedefop.europa.eu/files/6009_de.pdf.

Practical Tip

Collaborative Development of the Strategy Orientation

In many jurisdictions, the board of directors, executive board and management are responsible for defining the Strategy Orientation – and thus developing the mission, vision and strategic option. In the public sector, it is the responsibility of the legislator and the head of government. A collaborative development of the Strategy Orientation can take place through the active involvement of key experts. This approach is particularly useful because it integrates important information for decision-making. In the course of the «open strategy« development (Stadler et al. 2021), more and more organisations – from IT companies to the army – are involving external stakeholders to form an opinion on strategic questions and meaningful strategic answers. Studies show that heterogeneously staffed decision-making bodies are more effective (see Koryak et al. 2018).

We have listed below some helpful questions that the Strategility Master can ask Enablers and know-how providers to evaluate the strategic option.

Questions to assess the suitability of a strategic option

- Does the strategic option have an impact on the strategic challenge?
- Can the strategic option address the key risks in the external environment?
- Can the strategic option take advantage of the most important opportunities in the external environment?

Evaluation questions on the financial acceptability of the strategic option

- Is the ratio of estimated revenues to expenses satisfactory?
- Is the financial risk acceptable?
- Can we recommend this strategic option to our investors with a clear conscience?

Questions to assess the feasibility of the strategic option

- Can we implement this strategic option with the resources available or obtainable?
- Can we manage the implementation without compromising key processes and results?

The results of the evaluation can be visualised graphically using these three dimensions (see Figure 3.6); a template is provided in the *Digital Extras*. The axes represent the three evaluation criteria outlined above. You can choose your own scale, e.g. from zero to ten, with the higher number reflecting a better assessment. Your rating can be represented by a triangle for each direction of impact. Figure 3.6 shows two examples of strategic options. While the orange strategic option has clear advantages in terms of feasibility, the turquoise option is considered to be more appropriate and financially attractive. In this case, the Enabler would now need to balance feasibility against impact and performance.

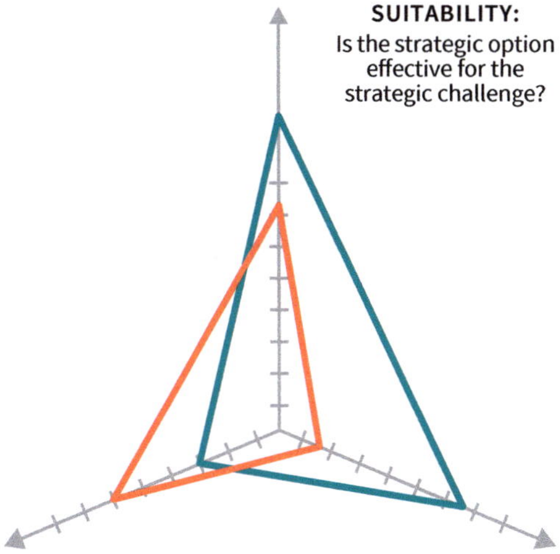

Fig. 3.6: Evaluation of strategic options

Event

Assumptions and Evaluation for an Effective Strategy Orientation

Objective: To become aware of the strategic challenge and clarify the impact on the direction of the organisation

Participants: Enabler, potential Strategy Master, potential Hypothesis Owner, Strategility Master

Work steps:

- Presentation of the initial strategic analysis (SWOT) and the resulting strategic challenge
- Making assumptions for an effective adaptation of mission, values, and vision, e.g. using the SWOT[3]. If necessary, initiate a process to adapt the mission statement.
- Proposing an effective strategic option and making assumptions about changes and strategic success explicit – for example with the SWOT[3]
- Transfer assumptions and corresponding indicators to be tested to hypothesis and strategy boards
- Evaluation of possible strategic options against the criteria of suitability, financial acceptability, and feasibility

Results:

- Assumptions and indicators of success-relevant developments in the environment are included in the hypothesis and Strategility Board.

- Assumptions and indicators about effective changes to the mission statement and strategic options are recorded in the hypothesis and Strategility Board.
- Decisions on adjustments to the mission statement
- Validated strategic options

3.5 From Strategy Orientation to Strategy Development

The validated strategic options form the transition to Strategy Development. The strategy concretises the strategic option. For the online newspaper, for example, the strategic option of market penetration would be specified with decisions on strategy elements such as products (in-depth articles on local, controversial topics), customers (left-wing educated middle classes in cities), and economic logic (subscription prices) (see Chapter 4).

Assumptions are also formulated and reviewed in agile Strategy Development – in the sub-process for effective strategy elements. You can also use these findings for the periodic impact assessment of the strategic option in the Review & Planning events. To this end, the Strategility Master records the assumptions from Strategy Development into the hypothesis board and the Strategility Board and arranges for them to be discussed with the Enabler in the periodic Review & Planning events. The Strategility process thus dovetails assumptions and insights from Strategy Orientation, Strategy Development, and Strategy Implementation (see Figure 3.1). This makes the Strategy Orientation sub-process iterative, as periodic discussions of current findings can lead to a readjustment of the orientation.

Practical Tip

Iteration in the Strategy Orientation

We recommend an iterative development of the Strategy Option, especially for the growth-oriented strategic options:

- Market penetration/service intensification
- Market/customer development
- Performance/product development
- Diversification

We do not recommend it for the consolidation-oriented strategic option of withdrawal. This is due to the negative impact and the negative reaction that a withdrawal from service areas can have on employees or business partners. In this context, it is unethical and damaging to the company's image to test prerequisites and effects.

By defining a strategic option to be pursued, the Enabler also determines the Strategy Owner, who is tasked with developing a concretised strategy for this strategic option. This process of concretising strategizing is explained in Chapter 4.

CHAPTER 4

Agility for Strategy Development

4 Agility for Strategy Development

Strategy Development is about finding an effective strategy that will help achieve the strategic option and thus answer the strategic challenge. Agile Strategy Development ensures that the impact of the selected strategy elements can be assessed. Strategically unfavourable decisions are identified at an early stage and the strategic option can be corrected if necessary before resources are invested in their implementation.

We have outlined the Strategy Development sub-process in Figure 4.1. The development of a more effective strategy starts with the validated strategic option from the upstream «Strategy Orientation« sub-process or the impact and results information from the downstream «Implementation« sub-process. To this end, the individual strategy elements and their underlying assumptions are collected and the most momentous ones are validated empirically or analytically. The result is a validated strategy. We explain this process in more detail below.

For your agile Strategy Development, you need to follow the five Strategility principles:

- Alignment with the strategic option and underlying mission statement (alignment)
- Making strategic assumptions explicit and validating the effectiveness of a strategy (evidence)
- A process that iterates through knowledge acquisition and decision-making (iteration)
- Both experts and decision-makers developing the basis for decision-making together (collaboration)
- Freedom of decision within defined guidelines for the people who drive the hypotheses and tests (autonomy)

In the following, we expand on the management explanations in Chapter 2 and provide you with specific information on how you can organise the management of Strategy Development in accordance with the five Strategility principles.

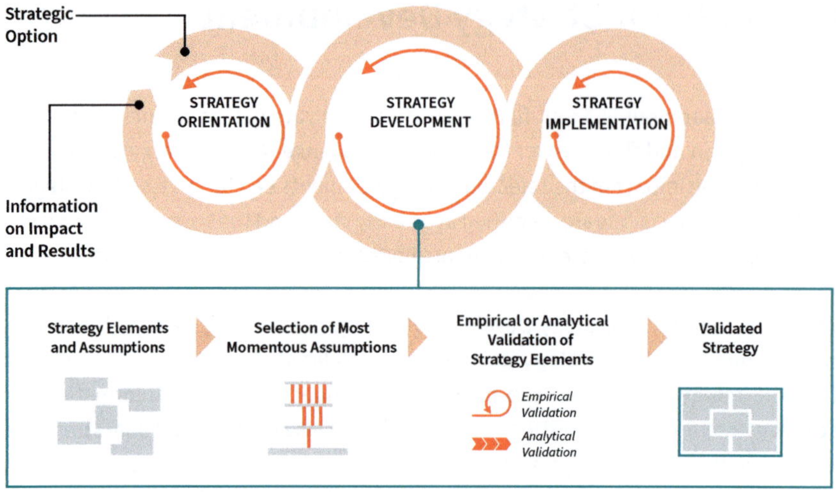

Fig. 4.1: Agile Strategy Development process

4.1 The Core: Strategies for Organisations and Administration

Structures for describing and visualising strategies have emerged in recent years[8]. In the following, we use the strategy elements based on Hambrick and Frederickson (2005) (see Figure 4.2):

1. The areas or playing fields in which an organisation operates – hereafter referred to as **arenas**. Five arenas need to be explicitly defined:
   a) The client and beneficiary groups targeted
   a) The products and services offered
   b) Sales territories and formats
   c) Core competencies that are important for good service delivery
   d) Value creation steps that the organisation performs itself – as opposed to those that it outsources
2. The **vehicles** used to develop the organisation. Here you can choose between:
   a) Internal, organic development
   b) Joint ventures/public-private partnerships
   c) Licensing, franchising
   d) Collaborations
   e) Mergers and acquisitions
   f) Participation in business ecosystems
3. The **differentiation** that distinguishes an organisation from others. Again, a choice can be made between:
   a) Price/cost efficiency
   b) Performance quality

8 Hambrick and Frederickson developed the «Five Elements of Strategy« for competitive strategies in 2005. The business model canvas by Osterwalder (2011) has become established for visualising business models.

c) Performance reliability
d) User friendliness
e) Image
f) Other factors

4. The strategic **timing aspects** of the development, such as specific time windows for a planned market entry or the launch of new service offerings
5. Finally, the **economic logic** by which an organisation generates profit – for example through economies of scale and low costs or through special services with higher prices. In public administration, this element of strategy is about deciding whether certain services should be provided particularly efficiently or whether additional expenditure on innovation, creativity or special proximity to citizens is acceptable.

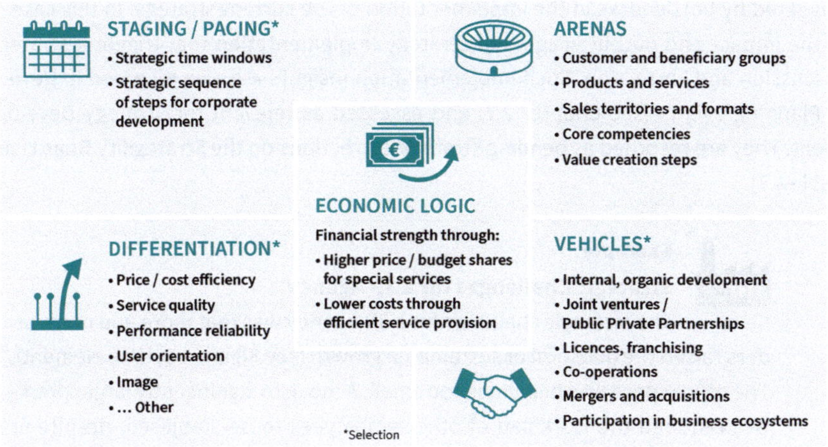

Fig. 4.2: Generic strategy canvas

These strategy elements should be explicitly defined and aligned with each other – including indirectly defining what is NOT part of the organisation's activities, such as beneficiary groups that are not the focus of the company's own activities. Over a period of one to three years, the strategy elements form a focal point for the organisation's activities and the use of resources. Their definition provides a common orientation for all organisational and administrative members in their work decisions.

In Figure 4.3, we have illustrated the strategy elements for a 3D agency that offers augmented reality visualisations of properties and products. In the *Digital Extras* you will find a canvas for the structured visualisation of your own strategy elements as well as an example for administration.

4.2 The Start: Acknowledging the Strategic Challenge

You will usually start with the strategic option defined in the previous process step – Strategy Orientation (see Chapter 2). For example, the strategic challenge might be that your administration is no longer able to invest, and the strategic option adopted for this could be: «We want to generate additional revenue by offering more attractive business relocation options.« This strategic option in product development is fed into the Strategility Board and triggers the development of service offerings, the definition of desired beneficiaries, and appropriate strategy elements in the Review & Planning event (see Chapter 2.7).

However, the discussion, definition, and harmonisation of strategy elements is often triggered by bottlenecks in the implementation of the current strategy. In this case, it is the impact and output insights of Strategy Implementation that trigger a strategy discussion and correction. Such implementation insights are also recorded in Review & Planning events (see Chapter 2.7) and assessed as relevant for Strategy Development. They are recorded as pending strategy corrections on the Strategility Board (see Table 4.1).

Example

Strategic Challenges for a 3D Agency

The strategic challenge for a 3D agency was that more and more orders raised the question of sustainable growth (see Figure 4.3, time element). The office space was becoming too small. A move to significantly larger premises required the workload of other employees to be stabilised, despite or alongside the volatile project business. This need brought new ideas for strategy elements: Could standardised 3D visualisations bring growth-enhancing consistency, and for which customer groups would they be attractive?
It was only in answering these questions that it became clear that the 3D agency's client base had not yet been defined, but was implicitly emerging. The challenge posed by the ongoing Strategy Implementation of achieving growth with stable capacity utilisation raised the question of a sustainable strategy. A Strategy Development programme was initiated.

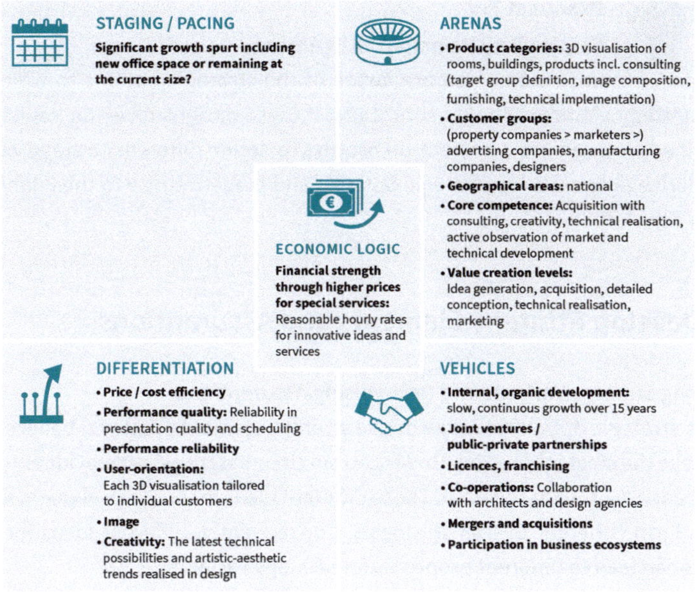

Fig. 4.3: Strategy content of a 3D agency

Sub-processes	Backlog	In preparation	In progress	Under observation	Completed
Orientation			Development of business growth and constant capacity utilisation		
Development	Develop strategy variant for growth with constant capacity utilisation		Determine roles: Strategy Owner, Hypothesis Owner, Strategy Exploration Teams		
Implementation	• Build a business case • Develop an project canvas				

Tab. 4.1: Strategility Board of a 3D agency – Start

Please also note the Strategility principle of alignment.

Practical Tip

Strategility Principle «Alignment«

Ensure that you are aware of the strategic options to which your strategy elements should contribute. If these options have not yet been defined, you can use a bottom-up process to define different strategy variants, derive proposals for strategic options, and present these to the Enablers for a decision.

4.3 Develop Strategy Elements and Assumptions

Developing Strategy Variants and Recognising Assumptions

From the strategic option, you now derive strategy elements that you believe will help you achieve the strategic option. Under certain circumstances, several ideas for a strategy element are brought together. These different ideas for strategy elements are combined to form coherent overall strategies. Consequently, different ideas for strategy elements can lead to different proposals for strategy variants.

Example

Strategy Variants for a Strategic Option

The 3D agency decided to respond to the strategic challenge of growth with constant capacity utilisation by focusing on the strategic option of product development.

One possible strategy variant for this strategic option focused on the new product of standardised 3D designs for the customer groups of furnishing office buildings and restaurant chains.

A second strategy variant for the strategic option of product development focused on the new product of 3D designs for integrated product and room presentations, that is, a combined product for the customer groups of real estate providers and furnishing manufacturers.

Suggestions for strategy elements are based on assumptions about their relevance and effectiveness. Some of these assumptions may have already been analysed in the initial analysis; others are rather unreflected. The Strategility principle of evidence suggests that assumptions should be formulated and the most momentous ones tested. Therefore, we recommend an approach that combines the development of possible strategy variants with reflection on the underlying assumptions. This work should be organised in collaboration with various experts. We suggest the following process:

Event

Strategy Variants and Assumptions

Objective: To develop appropriate strategies for the strategic option, and awareness of underlying assumptions

Participants: Strategility Masters, Enablers, potential Strategy and Hypothesis Owners, competent employees and managers

Work steps:

- Presentation of the strategic challenge and the strategic option (from the Strategy Orientation process)
- Questions: Which strategy elements make sense in order to realise this strategic option? What are our assumptions about why these strategy elements would be effective? Participants write their ideas on the assumptions canvas (see Figure 4.4 and *Digital Extras*).
- During the writing process, the Strategility Master guides the participants' thoughts towards the strategy elements by asking questions such as: What will change for our customers/beneficiaries if we implement this strategic option? Why do we think this will be effective? What will our value creation steps look like if we implement the strategic option? What are our assumptions about why this would make sense?
- The collected strategy elements are then combined into coherent strategies. Which strategy elements fit together to form a coherent strategy? The consistent strategy variants are added to the strategy canvas.

Output: Strategy variants for realising the strategic option and underlying assumptions

Further procedure:

- Presentation and evaluation of the strategy variants in terms of suitability, feasibility, and financial acceptability (see Chapter 3) with Enabler and Strategy Owner
- Decision on the strategy to be pursued by the Enabler
- Transfer of the strategy variants to be pursued to the backlog of the Strategility Board

The different strategy variants and their underlying assumptions are each mapped in a separate assumption canvas (see Figure 4.4). This canvas can also be found in the *Digital Extras*. As far as possible, the strategy variants should be assessed in terms of their suitability, feasibility, and financial acceptability, so that the Enabler, as the strategic decision-maker, can determine which of the strategy variants should be pursued further.

4

MISSION	VISION	STRATEGIC CHALLENGE
We stage messages with virtual reality solutions and make architecture, design and light approachable.	In a rapidly developing 3D world, we are constantly growing through expertise and innovation in value-added projects.	Achieve growt while maintaining capacity utilisation.

Strategic option to meet the challenge:
Product development for constant capacity utilisation

Possible strategy elements for implementing the strategic option "The following strategy elements are useful ..."		Assumptions about their effectiveness "... because presumably ..."
ARENAS		
Customer and beneficiary groups	Specialised furnishers of offices and restaurant chains	• 3D designs are the same for 90 % of the features • Standard 3D designs are desirable to make work easier
Product and service	Standard 3D designs	Marginal customisation effort allows multiplication
Sales territory and format	National	International competition too cheap
DIFFERENTIATION		
Price, quality, reliability, user orientation, image, ...	Quickly retrievable 3D designs	Previously unsatisfied need for fast 3D views
ECONOMIC LOGIC		
High prices, low costs	Medium price segment due to good quality but low costs	There are at least 50 quality-conscious standard suppliers

Fig. 4.4: Assumption canvas of a 3D agency

Practical Tip

Strategility Principle «Collaboration«

In our Strategility workshops, we have found that people who introduced a strategic change idea were unable to make significant progress with its structured testing. They felt that the search for consistent strategy elements and meaningful hypotheses and tests was unnecessary, time-consuming or even destructive. In organisational psychology, this phenomenon is described in terms of different roles in management (see Belbin 2004). We recommend that this phenomenon be consciously managed: An energetic, enthusiastic idea provider may not enjoy or be able to systematically search for evidence for their idea. This could be carried out by the Hypothesis Owner; the idea provider may then only become involved again when the test results are discussed.

Defining Strategic Hypotheses

For the strategy variants that will be pursued further, identify the momentous assumptions (see Chapter 8). These are the assumptions that, if disproved, would invalidate all other activities – in the case of the 3D agency, for example, the assumption that standard 3D designs are desirable for specialised office fitters. You then carry out a triage for these momentous assumptions:

- Which assumptions already have sufficient information?
- Which assumptions are «just complicated« and can be analysed with expertise and experience?
- Which assumptions are complex and need to be tested empirically with hypotheses and tests?

Figure 4.5 illustrates the development of strategic hypotheses.

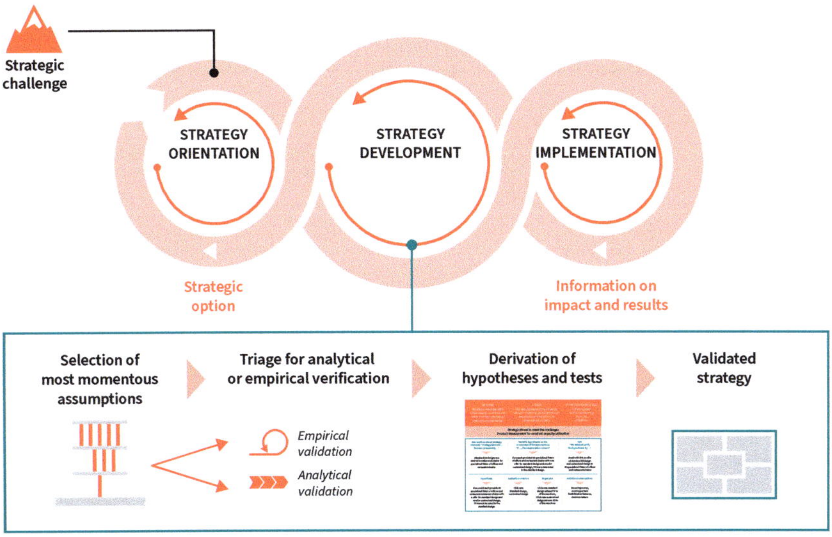

Fig. 4.5: Development of strategic hypotheses

The assumptions to be empirically tested can be tested and assessed using an agile, iterative process. Testable hypotheses are formulated for these assumptions (see Chapter 8.5). For example, testable hypotheses could be formulated for the assumption that standardised design concepts are desired by office and restaurant fitters:

- If we approach 50 specialised office and restaurant fitters with an offer for a standard design and an offer for a customised design, 75% will be interested in the standard design.
- If we need less than 10 hours to adapt standard 3D designs to new applications or customers, then designs can be standardised efficiently.

In the 3D agency, the momentous assumption, that standard designs are desired to make work easier, was concretised with three ideas for testable hypotheses (see Figure 4.6). A corresponding hypothesis canvas can be found in the *Digital Extras*. The first hypothesis was chosen because it promises valid feedback on the interest of the fitters with little effort. The task of testing this hypothesis is noted in the backlog of the Strategility Board (see Table 4.2).

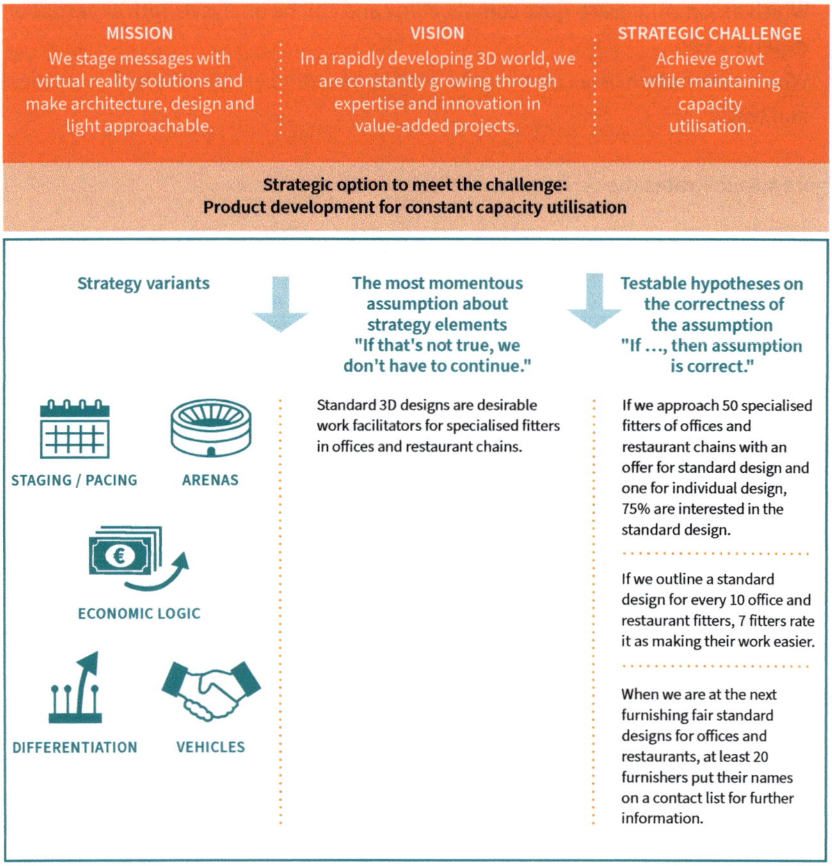

Fig. 4.6: Strategility hypothesis canvas of a 3D agency

Sub-processes	Backlog	In preparation	In progress	Under observation	Completed
Orientation				Development of business growth and constant capacity utilisation	
Development	Test the hypothesis, that office and restaurant fitters are interested in standardised 3D design		Sketch test		• Develop strategy variant for growth with constant capacity utilisation • Define roles: Strategy Owner, Hypothesis Owner, Strategy Exploration Teams
Implementation	• Build a business case • Develop a project canvas				

Tab. 4.2: Strategility Board of a 3D agency – hypothesis

At this point we outline an event for formulating and selecting the hypotheses to be tested.

Event

Definition and Selection of Strategic Hypotheses

Goal: Testable strategic hypotheses that provide meaningful information about the effectiveness of strategy elements

Participants: Strategility Masters, potential Hypothesis Owners, Strategy Exploration Team

Work steps:

- Presentation of the strategic challenge, the strategic option and the strategy variants, the latter illustrated using the strategy canvas (see Figure 4.2 and *Digital Extras*)
- Explanation of the assumptions made
- Identification of the momentous assumptions

- Triage for the momentous assumptions: Which are sufficiently certain, which need analytical validation, which need empirical testing?
- Formulate hypotheses for assumptions to be tested empirically. Possible assignment: «Formulate if-then sentences for the assumptions. For example: If we offer xy, then the partners will react as follows – and this confirms our assumption.« Participants write their if-then sentences on moderation cards.
- Clustering of hypotheses with similar content
- Selection of the hypotheses that allow the most significant increase in knowledge (see Chapter 8)
- Transfer of the hypotheses to be tested to the backlog of the Strategility Board

Output: Collection of hypotheses to be tested in an agile manner; selection of the most meaningful hypotheses with which to start testing

Further procedure: Design of tests for the hypotheses

4.4 Testing Strategic Hypotheses

Agile Strategy Development takes place in an iterative sequence of knowledge acquisition and decision-making.

Designing and Carrying out Strategy Tests

You design and implement the first tests of the defined hypotheses. Recommendations and tools for test design can be found in Chapter 8. To manage the test design and implementation, we recommend the following:

- **One test for multiple hypotheses:** Brainstorm ideas and consider whether you can test multiple hypotheses with one test. In the 3D agency, for example, it may be possible to conduct 20 customer meetings with the outlined content of a standard design in a kind of «fake door testing« (see Chapter 8.6) and note both whether a customer need exists and what adjustments would be necessary from customer to customer, and then calculate the internal effort this would entail.
- **Indicators for the test evaluation:** Ask the Hypothesis Owner and the Strategy Exploration Team to define indicators for the test procedures, which you can use to judge whether your hypothesis should be accepted or rejected. These indicators will be used to evaluate the test results. For example, if the indicator for the 3D agency is «out of 20 customers approached, at least 15 would like to be informed further about our range of services«, the question could be asked at the end of a customer meeting as to whether the dialogue partners would like to sign up to a list for further information.

Practical Tip

Strategility Principle «Evidence«

Explain repeatedly to the Strategy Exploration Teams that the aim is to gather information with an open mind, even if it is information that disproves the key hypotheses. It is a human impulse to seek confirmation for key assumptions. Strategy Exploration Teams need to be aware that their findings should be as unbiased as possible to serve as an informational basis for decision-making. Therefore, the goal is not getting participants to sign up for further information about the range of services at the end of a customer discussion, but rather leaving them with a very open choice.

- **Determine the number of tests:** Similar to the indicators that you define for the test evaluation, you should also clarify from the outset how many tests you need to evaluate the test results. This decision is made by the Strategy Exploration Team together with the Hypothesis Owner.
- **Determine the number of iterations:** Determine how many test sessions and adjustments you consider acceptable. This decision is related to the available resources such as capacity, time, and budget. The decision is made by the Strategy Exploration Team together with the Hypothesis Owner.
- **Use testing as an opportunity to communicate:** Think of testing not only as an opportunity to confirm or disprove a hypothesis, but also as an opportunity to communicate with stakeholders. For example, when talking to customers about standard 3D designs, you may find that people sometimes feel sick when taking a virtual tour of a property because their movements are too erratic or too fast. This gives you unexpected clues about the design of your offering – in this case, providing a virtual person to guide visitors through the rooms slowly and with steady movements.
- **Document the tests and experiences:** Especially when several Strategy Exploration Teams are working in parallel, it is important to document experiences of test progress and findings so that they can then be systematically shared with each other. This will enable you to adapt the test procedures if necessary. Without documentation, important sources of knowledge can be lost, such as the fact that the interior designers of municipal hospitals present the 3D visualisations of the properties to a committee of participants who are usually inexperienced in real estate. In contrast, the organisers of outpatient and inpatient hospitals tend to delegate clarification to facility managers with real estate experience – important information for the possible formation of customer and service design subgroups.
- **Adapt test procedures:** Our Strategility work has shown that test procedures should be adapted after three to four administrations. There may be two reasons for this:
  - Initial results become clear, which can be specified in the further test runs. For example, after four test interviews, it becomes clear that hospital designers pay particular attention to 3D visualisations of the common areas. In the subsequent

4

test interviews, it is then of added value to specifically address these common areas. The content of the test procedure should be adapted accordingly.

- The test procedure is not powerful enough: During the first few test administrations, it may become clear that your test respondents do not fully understand the content, for example, that you need to ask them to make a decision or take an action in order to generate meaningful tests, or that the test procedure needs a different structure.

Practical Tip

Strategility Principle «Iteration«

The strength of the statement is tested with the following three questions:

- Are we receiving statements from the people we are interested in, or are we only reaching indirect representatives? If so, the test group must be changed.
- Can we really answer our hypothesis with the outputs, or are we missing the crucial information for an assessment? If so, the test procedure needs to be changed.
- Are the tests really neutral and independent of the respective test situation, or would a different Strategy Exploration Team achieve different results? If so, the test procedure needs to be designed to be more neutral.

For the ongoing test assessment, retrospectives after two to four test runs are recommended in order to make adjustments. Especially if there are several Strategy Exploration Teams, it is worth exchanging information about content and process findings and the need for adjustments. You will find content and methodological hypothesis evaluation canvases in the *Digital Extras*, as well as an explanation in Chapter 8.8.

Practical Tip

Strategility Principle «Autonomy«

Especially for larger strategy projects with multiple tests and Strategy Exploration Teams, the teams should be able to decide independently on test procedures and processes, as otherwise information gathering will be delayed by feedback from decision-makers. To this end, it must be clear to the Strategy Exploration Teams what information is being sought on hypotheses, strategy elements, and strategic options. They also need to be clear from the beginning about the authority they have to carry out testing. Hypothesis Owners and Enablers should agree on this.

You can record your test planning in the test canvas (see Figure 4.7). You can also find this canvas in the *Digital Extras*.

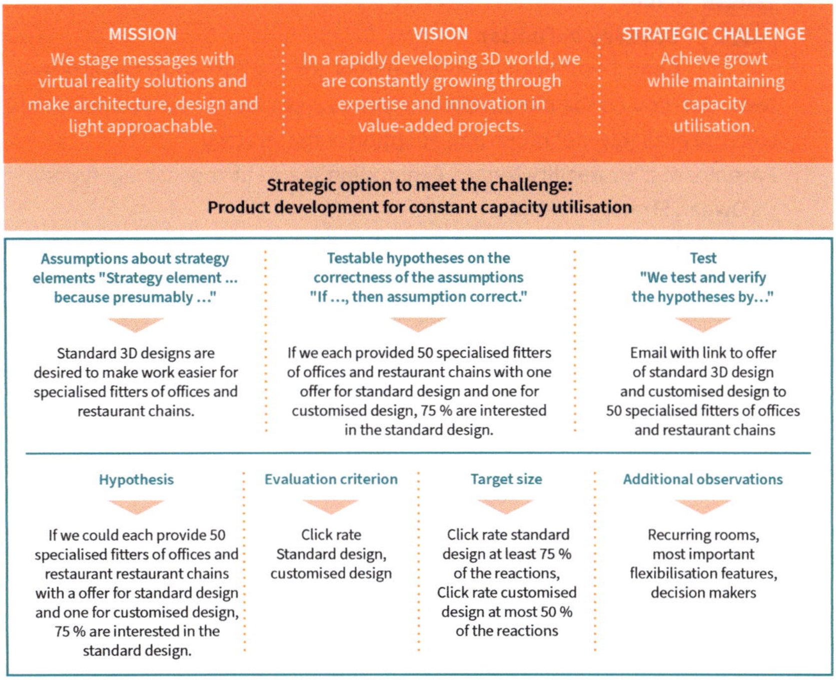

Fig. 4.7: Strategility test canvas of a 3D agency

4.5 Validate Strategic Hypotheses

Strategy tests are organised in strategy sprints of approximately two weeks and evaluated in sprint-specific Review & Planning events with regard to the significance and content of the findings (see Chapter 2.5). The test teams explain their findings using the test evaluation canvas (see *Digital Extras*). For the final assessment of the strength of the statements, it is important to understand and organise the evaluation not as an expression of the personal skills of the Exploration Teams, but rather as decision information. Make it clear to the Strategy Exploration Teams that it is a strength to point out a lack of conclusive evidence. The Hypothesis Owner decides together with the Exploration Team how to proceed with the hypotheses and tests.

Periodically – approximately quarterly – the Strategility Master contributes the consolidated results and findings from the sprint reviews to the overarching Strategility Review. The hypothesis and Strategy Owners analyse the statements on the strategic assumptions and the preferred strategy elements and then decide on how to proceed: Do the findings fundamentally confirm or refute previous strategic decisions? What adjustments need to be made to the strategic decisions? If the findings are sufficiently confirmatory, the strategy can be defined and approved for implementation.

Event

Strategy Definition

Goal: Feedback and evaluation of the outputs from the strategic tests; decision on the effectiveness of the strategy elements; if sufficient results are available, definition and adoption of the strategy

Participants: Strategility Master, Exploration Teams (if applicable), Hypothesis Owner, Strategy Owner

Work steps:

- The Hypothesis Owner or the Exploration Teams present the consolidated test results using the hypothesis canvas.
- Everyone discusses the significance of the test results for the strategic assumptions and the effectiveness of the strategy elements in addressing the strategic challenge.
- The Strategy Owner decides whether to retain or change the overall strategy elements and the strategy.
- The Strategy Owner decides whether the momentous assumptions of the strategy have been confirmed and whether they can assess their impact well enough to release them for the initialisation of implementation (see Chapter 4.6).

Outputs: Hypotheses have been evaluated; assumptions and strategy elements have been evaluated, the overall strategy has been developed and, if necessary, approved for implementation by the Enabler

To manage strategy validation, we recommend that you consider the following aspects:

- **Decision-making authority regarding evaluation**
  In order to control the testing and hypothesis evaluation, you need to clarify the decision-making authority. It goes without saying that this allocation of authority must match your management style. As an example, we suggest the following division:
  - The Strategy Exploration Team decides on the test procedure, the test iterations, the informative value of the test procedure and the need to adjust the test procedure.
  - The Hypothesis Owner decides on the meaningfulness, test termination, and acceptance or rejection of the hypothesis.
  - The Strategy Owner decides whether to accept or adapt the strategy elements.
  - The Enabler decides on the strategy variant.
- **Content evaluation of the hypothesis**
  Together with the Hypothesis Owner, it is necessary to analyse whether the test results indicate acceptance or rejection of the hypothesis. according to the defined indicators. Again, it is important to emphasise that this assessment does not represent an evaluation of the individual abilities of the Hypothesis Owner, but merely information that should form the basis of an objective decision.

- **Content evaluation of the strategy element and the overall strategy**
  The significance of a hypothesis for the strategy decision is assessed on the basis of its meaning and content (see Chapter 8.7). The Strategy Owner reflects in the Review & Planning event: Are the insights sufficient to decide on a strategy element?
  - If serious hypotheses have been credibly refuted in tests, the Strategy Owner must decide whether and how the relevant strategy element can be adjusted in order to continue to pursue the overall strategy. For example, if the 3D agency finds that there is too little customer interest in standard designs, the overall strategy around standard designs must be abandoned. An alternative strategy must be found to manage the direction of growth, for example by reducing the variety of products offered and specialising in certain designs.
  - If the momentous hypotheses are confirmed and the secondary hypotheses are refuted, the Strategy Owner should consider under what circumstances the strategy element in question can still have a positive impact on the overall strategy. For example, if the 3D agency disproves that the effort required to customise standardised offerings is satisfactorily low, other strategy elements can be adjusted if necessary. For example, the economic logic can be changed from low prices to higher prices. To assess these new strategy elements, new hypotheses and tests may be needed, which will be added to the backlog of the Strategility Board.

To keep track of hypotheses and tests, you can work with a hypothesis board, which lists all hypotheses and tests with their results. An outline is given in Table 4.3.

	Backlog	**In progress**	**Accepted**	**Refuted**
Assumptions	Assumption 4	Assumption 3	Assumption 2	Assumption 1
Hypotheses		Ass3 hypothesis 3	• Ass2 hypothesis 1 + 2 • Ass3 hypothesis 2	• Ass1 hypothesis 1-3 • Ass3 hypothesis 1
Tests	Ass3-Hyp3 test2	Ass3-Hyp3 test1		

Tab. 4.3: Hypothesis board

Practical Tip
Strategility Principle «Evidence«
It is possible that the decision-makers continue to pursue a strategy even though the corresponding hypotheses have been refuted. This may be due to intuition or general organisational developments. Decision-makers in an organisation bear the decision and development risk and must have this freedom of choice. To avoid damaging the agile decision-making culture, it is important that these decisions are made and justified explicitly, rather than being made implicitly.

- **Understanding hypothesis evaluation objectively as a basis for decision-making**
  As with test evaluation, neutrality, and objectivity are also important when assessing hypotheses. Neither acceptance nor rejection of a hypothesis should be seen as a personal success or failure. Pre-formulated sentences may be helpful for objective evaluation, e.g.: «The hypothesis test was a success because we have learnt that ... This allows us to reject/accept the hypothesis with good reason.«

4.6 Transition to Implementation

Based on the analysis and tests carried out, the Strategy Owner can assess when a strategy is sufficiently validated to be implemented. This may already be the case once the momentous assumptions have been confirmed. If the 3D agency has identified an interest in country-specific standard 3D designs among restaurant fitters, it can implement this strategy variant. Assumptions about individual strategy elements may not yet have been validated, such as changes to the company's own value creation steps or the design of the economic logic. Since the Strategility approach involves a constant flow of insights, it is possible and useful to carry out these clarifications through agile implementation projects. To this end, the unresolved assumptions on strategy elements are recorded in the implementation backlog of the Strategility Board (see Table 4.2).

The Strategy Owner instructs the person responsible for implementation to drive implementation forward and, in turn, to report back the results for strategy validation in the Review & Planning events. This feedback flow is controlled by the Strategility Master by continuously updating the Strategility Board and the hypothesis board with the outputs, and planning Review & Planning events to feed back the implementation results into the Strategy Orientation (see Chapter 5.3).

CHAPTER 5

Agile Strategy Implementation

5 Agile Strategy Implementation

5.1 Overview

Strategility flow links Strategy Orientation, Strategy Development, and Strategy Implementation. By involving knowledge holders in the Strategy Development process, the strategy is embedded in the organisation or administration at an early stage, not just during Strategy Implementation. This significantly reduces the risk of lacking strategy acceptance during implementation.

When implementing the strategy, the aim is to achieve the optimal impact of the strategy. Agile Strategy Implementation ensures that strategy elements can be adjusted or the strategic option can be fundamentally corrected if new insights emerge. In order to be able to react quickly, the information about the impact of the strategy from the initiatives must be fed back into the Strategy Development continuously. Organising the evaluation and feedback into the Strategy Development process is a task of agile Strategy Implementation steering.

With agile Strategy Implementation you get:

- Clarity about the strategic option, strategy, and strategy elements and their intended impact (alignment)
- Planning implementation controlling BEFORE implementation in order to effectively determine impact, output, and process information on ongoing implementation projects (evidence)
- Continuous feedback of Strategy Implementation into Strategy Development in order to be able to assess the effectiveness and feasibility of implementation projects (iteration)
- Exchange between Strategy Owners and Implementation Teams to assess the impact of implementation (collaboration)
- Freedom of decision for the Implementation Teams to achieve the optimum impact from the implementation projects (autonomy)

You will now receive an overview of agile Strategy Implementation. Each element is then described with examples and tools in the following sub-chapters.

Implementation initialisation lays the foundation for successful Strategy Implementation. This involves creating the strategy roadmap and business case which are drawn up. These provide information on the timeline, costs, and resources required to implement the strategy. In administration, the business case typically corresponds to the cost-benefit analysis. At the end of the implementation initialisation, the Enabler, with the involvement of the Strategy Owner, decides whether to approve the Strategy Implementation initiative. We know from numerous projects that organisations and administration can be overwhelmed by Strategy Implementation because their activities are focused on the operational business – not on developing the business. Therefore, before the Enabler decides to approve the strategy, it is necessary to check whether the conditions for successful Strategy Implementation are met.

Figure 5.1 shows that Strategy Implementation often involves a number of projects that are closely interrelated and need to be coordinated accordingly. The projects can be implemented using both agile and plan-driven methods. Multi-project management methods (portfolio management, programme management and project management) and agile development methods (Scrum, SAFe®, KANBAN etc.) are commonly used for this purpose.

In practice, we often see that strategy controlling only takes place after the strategy has been implemented. Obviously, there is a significant risk that it will be too late to react to new findings and that change costs will be unnecessarily high due to unfavourable decisions. For this reason, controlling activities are already planned when the implementation initiative is defined. This ensures that you are aware of relevant information and can process it.

Roles in Strategy Implementation

Strategy Implementation also requires clear roles and responsibilities. The Strategility Master is responsible for the organisation of implementation initialisation and implementation controlling. The roles of the person responsible for implementation and the Implementation Team depend on the chosen implementation method, which is why we use these generic terms instead of roles such as project manager, portfolio manager, etc. Chapters 5.6 and 5.7 provide information on how the roles could be organised depending on the method.

Agile Strategy Implementation consists of several elements, which you will learn about in more detail in this chapter:

- Initialisation of Strategy Implementation
- Setting the stage for successful Strategy Implementation
- Agile implementation controlling
- The definition of implementation methods and their agile use for Strategy Implementation

The examples of how different methods (HERMES, SAFe®, OKR) can be used for agile Strategy Implementation will deepen your knowledge and support you in applying Strategility.

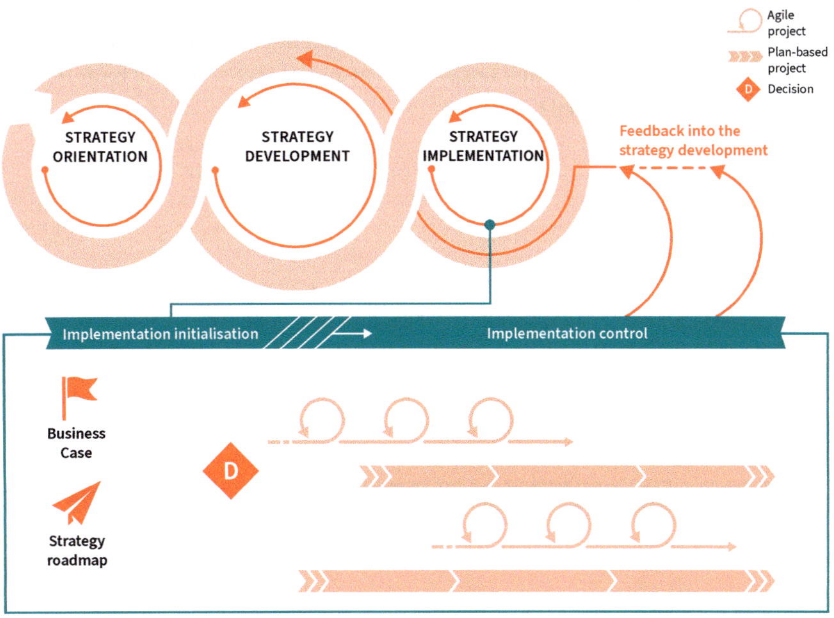

Fig. 5.1: Strategy Implementation with implementation initialisation and implementation controlling

5.2 The Initialisation of Strategy Implementation

In the initialisation phase, the implementation is structured and planned so that the implementation initiatives, the implementation organisation and the implementation controlling are defined. The capabilities of the organisation or administration to successfully implement the strategy are also analysed and the prerequisites are created or planned. At the end of initialisation, the expected effectiveness of the strategy is reviewed and the Enabler decides on the release of the Strategy Implementation with the involvement of the Strategy Owner.

Implementation initialisation is based on the results of Strategy Orientation and Strategy Development:

- The strategic challenge and the chosen strategic option
- The tested and selected strategy elements with their assumptions and hypotheses

Based on these principles, the strategy roadmap and the business case are developed (see Figure 5.2):

- The roadmap provides an overview of the process with a timeline and the learning points (milestones) at which the impact will be reviewed.
- The business case refines the planning and estimates the economic effectiveness of the strategy. Typical questions are: What are the costs of implementation and ongoing operations? What benefits/revenues will be generated over what period of

time? How will it be financed? What human resources will be required for the implementation initiatives? What specific expertise will need to be developed? What is the broad outline of the future organisational and operational structure? How will sustainability be achieved?

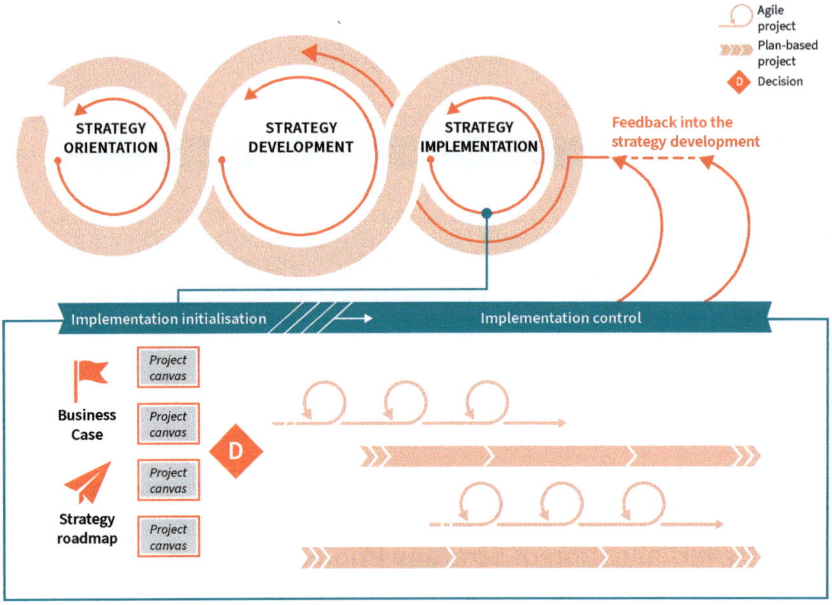

Fig. 5.2: Outputs of the implementation initialisation

If you work in administration, the focus may be on the benefits rather than the returns. In this case, you may describe the business case as more of a cost-benefit analysis, but you should still be able to answer the questions listed above.

The names of the strategy roadmap and the business case, as well as the exact content, may vary depending on the implementation method chosen. However, the outputs should cover the topics described above.

To ensure that the roadmap and the business case are sufficiently reliable as a basis for the decision to implement the strategy, the specific implementation initiative must be identified and defined. The roadmap, business case, and project canvas are preferably developed iteratively so that they are aligned and realistic at the end of the initialisation. Realistic means that the uncertainties are identified and lie within a range that is acceptable to the Enabler.

To carry out the initialisation, the Strategy Owner appoints a person who is responsible for developing the results. Depending on the situation, this person will involve people in the role of the Implementation Team.

As a prerequisite for impact-oriented implementation controlling, an project canvas is created for each implementation project, showing, among other things, when which impact is to be achieved. The example in Figure 5.3 shows an project canvas for a company in the pressure sensor industry that wants to provide engineering services in the USA directly on site at the customer's premises instead of only from Switzerland as before.

BRIEF DESCRIPTION OF THE PROJECT
Engineering services are provided in the USA by engineers directly on site at the customer's premises.

ORGANISATION / RESPONSIBILITY
Strategy owner: CEO USA
Responsible for implementation: Head of Marketing USA
Involved in implementation: Head of Engineering

KEY OUTPUTS / MILESTONES
Staff available: 1.5.20nn
Processes adapted: 1.7.20nn

IMPACT / TIME
10 orders acquired: 1.3.20nn
Cost reduction of 20%: 1.8.20nn

HYPOTHESES / ASSUMPTIONS
Customers accept Engineers
Customers see on-site work as an advantage
Work permits for USA are granted

GENERAL CONDITIONS / RESTRICTIONS
Market area USA: California and Texas

EXPENDITURE / COSTS
Internal personnel expenses: 20 PT
Costs: € 20,000

Fig. 5.3: Example of an project canvas

In the *Digital Extras* you will find a canvas to describe your Strategy Implementation projects.

> **Practical Tip**
> **Ready to Make the Decision to Implement the Strategy?**
> Before the Enabler makes the decision to implement the strategy, a defined level of maturity should be reached:
>
> - The roadmap and business case or cost-benefit analysis are available with an acceptable level of planning accuracy. They confirm the chosen strategic option and strategy elements.
> - The hypotheses for the momentous assumptions have been defined and tested during Strategy Development. Assumptions that have not yet been tested are recorded in the project canvas. This means that they can be continuously reviewed in the light of new findings during the implementation of the initiatives.
> - A complete canvas is available for each strategic project. If new insights later mean that further projects are required or ongoing projects need to be adapted, the project canvas is also created or updated.
> - A rough assessment regarding feasibility (technical, organisational, etc.) has been carried out, and boundary conditions and unalterable constraints should be known. This may require that the first projects

are already partially or fully implemented during the initialisation of the Strategy Implementation.

- The methods for implementation have been determined; implementation controlling and the learning points for impact measurement have been defined.
- The organisation is ready to implement the strategy. Measures to close identified gaps have been defined and their implementation has been ensured (see the last part of Chapter 5.4).

The criteria for the approval of the Strategy Implementation can be found as a checklist in the *Digital Extras*.

5.3 Agile Implementation Controlling

Agile Strategy Implementation controlling deliberately defines learning points at which the impact is reviewed. This involves analysing what impact was planned and what was achieved. This learning-based process feeds back into Strategy Development. This enables the Strategy Owner to monitor impact achievement and, if necessary, adjust strategy elements and the strategy as a whole – or consult with the Enabler about adjusting the strategic option.

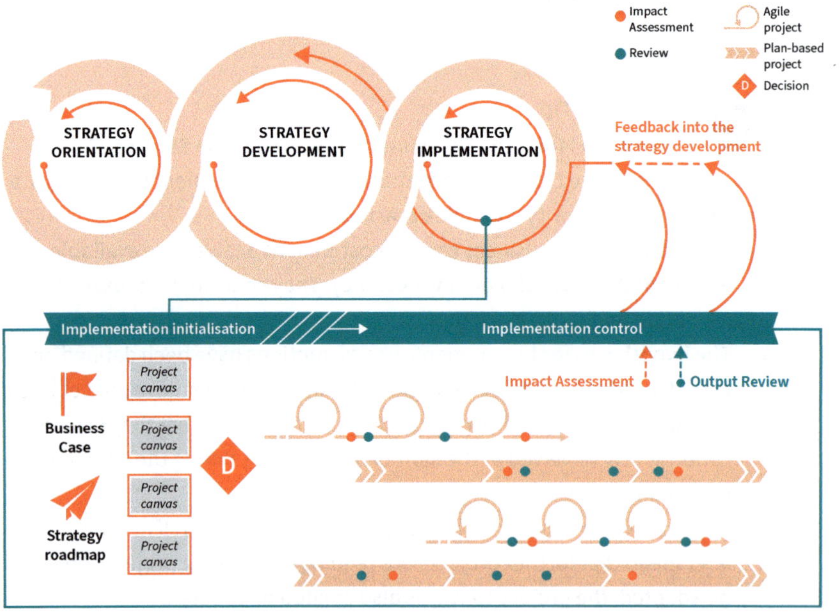

Fig. 5.4: Implementation controlling

In addition to impact measurement, output measurements are also carried out.

The Distinction Between Output and Impact

«However beautiful the strategy, you should occasionally look at the results.« This quote, which is probably misattributed to Winston Churchill, leads to the question of what the result of the strategy is. For clarity, we distinguish between output and impact.

For example, if we consider «market penetration« to be the strategic option and the manufacturer of pressure sensors builds a new logistics centre in order to be able to supply its customers more quickly and thus retain them, the new building is an output of the Strategy Implementation. Whether the organisation will be successful with the new logistics centre and whether the impact of fast delivery and customer retention will occur, cannot be said with the completion of the logistics centre.

Example

Output and Impact

The manufacturer of pressure measurement sensors wanted to better serve its US market by increasingly providing engineering services on site at the customer's premises in the future. It has defined the following strategy elements, among others:

The strategy elements (excerpt)

- Customer segment: Water production and distribution companies
- Geographical dimension: California, Texas
- Product category: Engineering of pressure sensors for water treatment
- Economic logic: Lower costs through customer proximity and simpler processes for the same price
- Business development methods: Services provided by two in-house engineers who relocated from Switzerland to the USA

The desired financial impact

When analysing the business case for this practical example, we focus on the strategic evaluation criterion of financial acceptability and assume the following impact:

- Economic logic: Lower costs through customer proximity and simpler processes for the same price → **Cost reduction of 20% leads to a margin of xy per order.**
- Product category: Engineering → within 6 months, we are able to acquire **10 additional orders for engineering services.** This leads to an **increase in turnover of € 200,000** in the first half of the year.

Figure 5.5 illustrates the impact measurement based on the practical example at different points in the course of the Strategy Implementation initiative. The feedback to the Strategy Development makes it possible to check whether a strategy adjustment is necessary.

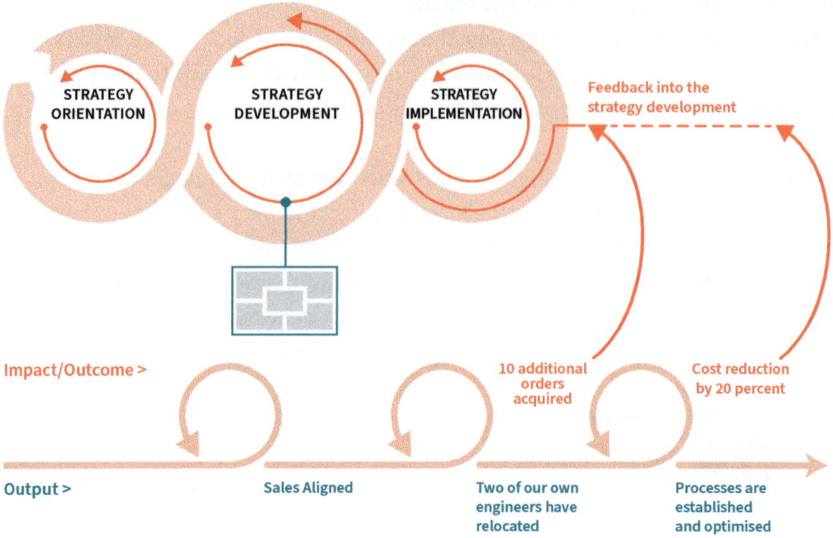

Fig. 5.5: Continuous impact measurement and feedback into Strategy Development

Example

Digitisation with an Unexpected Impact

A Municipality in the Zurich Area replaced its municipal administration software with a new generation product. At the same time, the process for managing the town hall was digitised.

Before digitisation:

Local associations were able to use the community hall free of charge. An email was sent to the hall manager to reserve the room. If the hall was free on the planned date, the reservation was confirmed by email. If the hall was occupied, further communication was made to find a suitable date. The administrator kept a schedule of room bookings. Once the booking had been placed, no further steps were necessary.

After digitisation:

Once an account has been created, room availability can be checked online and a provisional room booking is required. An application must then be printed out, duly signed and stamped and sent by post to the municipal administration (an electronic signature is not provided for the time being, but alternatively the application can be scanned and emailed to the administration on request). Once the application has been checked, the municipal administration will send a signed rental agreement by post. This must be countersigned and returned by the tenant. This completes the reservation. The municipal administration will update this manually in the software. After the event has taken place, an acceptance report, which can be downloaded from the municipality's website, must be completed, printed and returned (by post or scanned and sent by email). If there are no objections, this will be signed and returned by the municipal administration. As the new software cannot differentiate between residents' associations and other tenants, the municipal regulations have been adapted and the municipality will now initiate the invoicing process. The invoice is printed by the administration and sent to the tenant by post. After receiving the invoice, the tenant transfers the new payment of CHF 50 via e-banking. Finally, the payment must be entered manually in the association's accounts.

Conclusion:

The project for the introduction of the new municipal software was carried out successfully, i.e. on time and within budget. However, the impact is very negative from the point of view of the administration's customers.

The distinction between output and impact has several advantages:

1. Responsibility can be clearly assigned:
   - The Enabler is responsible for the strategic option and bears the associated consequences.
   - The Strategy Owner is responsible for the impact of the strategy elements.
   - The person responsible for implementation and the Implementation Team are responsible for the output of the Strategy Implementation.
2. A distinction can be made between checking/accepting the output and measuring the impact. The different procedures ensure clear responsibilities:
   - The person responsible for implementation and the Implementation Team are responsible for checking/accepting the outputs.
   - The Strategility Master is responsible for ensuring that impact is measured and fed back into the Strategy Development process.
3. Requirements from the desired impact flow into the output as requirements:
   - Whether an impact is sustainable also depends on the output. For example, if the logistics warehouse is to be able to adapt flexibly to new logistical requirements, the corresponding requirement must be included in the warehouse design.
   - Requirements for sustainability or environmental compatibility can be clearly assigned to the output.

The Implementation Controling Process

Implementation controlling is a continuous process. It consists of:

- The controlling of the three variables: impact, output, and procedure
- The assessment of deviations
- The distribution of information
- Ensuring that decisions are made

The results of Strategy Development, the business case, the roadmap, and the project canvas form the basis for implementation controlling and the assessment of impact,

5

output, and approach. The «Check & Adapt Strategy Implementation« event is used to assess the impact/output/procedure. This is carried out for each initiative. It can be event-based – for example at learning points – or periodic. The work steps are adjusted accordingly.

Event

Check & Adapt the Strategy Implementation

Goal: Examination of the impact, output, and procedure of an implementation initiative leads to insights and optimisation

Participants: Strategy Owner, Implementation owner, Implementation Team

Work steps:

- **Brief presentation:** State of the projects with learning points achieved, status of outputs, effort, and costs
- **Output assessment:** Questions are: Do the outputs meet their purpose? Do they meet the requirements? Are there new requirements or adjustments to the requirements? Do the outputs comply with the organisation's specifications and constraints? Are the outputs effective?
- **Impact assessment** at learning points: Questions are: What impacts have we been able to check? What did we learn? Which assumptions from the Strategy Development have been confirmed or refuted? What impact optimisations are possible? What measures are necessary for this? Do strategy elements need to be adjusted? Is the business case achievable and is the strategic option still appropriate?
- **Procedure assessment** with regard to effort/costs/time/resources compared to the project canvas, business case, and roadmap: Questions are: Where do we stand and where are there deviations? Does the chosen approach meet the stakeholders' requirements? How is the risk assessment? What is going well, what have we learnt and where can we improve? What improvement measures are appropriate?
- Evaluate and decide on the measures identified from the impact/output/procedure assessment

Output:

- Insights (learnings)
- Decisions on measures and further implementation

Further procedure:

- Follow-up of the outputs in the Strategility Board by the Strategility Master and/or the person responsible for implementation
- Inform Enablers about key insights

Practical Tip

Autonomous Implementation Teams and Customisation of Outputs

In principle, the Implementation Team has the expertise to adapt the outputs. For example, instead of sending experts to the US, the pressure sensor manufacturer could hire and train local staff. The aim is always to achieve the optimum impact, such as minimising labour and process costs. Of course, the requirements of the organisation or administration must be met and sustainability ensured. If it turns out in the case study mentioned above that the two engineers from Switzerland are not achieving the required result and it would be better to use engineers from the USA, this adjustment should be made quickly. The relevant decision-making authority should be agreed explicitly. Depending on the project management method and the governance model of the organisation, this should be recorded in the role description or in the project canvas.

The Conclusion of the Strategy Implementation

Once the strategic challenge has been answered or resolved and all implementation projects have been completed, the Strategy Implementation is formally closed.

The formal conclusion of Strategy Implementation depends on the chosen method of Strategy Implementation and the governance requirements of the organisation or administration. For example, a final report is usually written and approved at the end of projects or programmes. As the effectiveness of the Strategy Implementation and the strategy itself are continually reviewed, this is not part of the final report.

5.4 Prerequisites for Agile Strategy Implementation

To ensure that Strategy Development and implementation are successful, it is worth taking a closer look at the success factors. If you have already been involved in such initiatives before, you can draw conclusions for the next initiative from your wealth of experience of previous successes and failures. We have therefore compiled some success factors from our own experience that we would like to share with you. Of course, the list could be extended at will.

Corporate culture is so important that we have dedicated a separate chapter (Chapter 7) to this topic. The agile principles underlying Strategility are not embedded in all organisations and need to be consciously addressed. Systematic learning from insights, managing for impact rather than outputs, and roles and their responsibilities are just a few examples that require a significant change of mindset in organisations and admin-

5

istrations. This cultural change can be supported by methods, but above all it requires leadership by example. In many cases, it is also advisable to create a change plan that consciously addresses the variety of change elements. The experienced Strategility Master always takes the cultural dimension into account and optimises the process accordingly.

The resource situation is an ongoing issue. Many organisations and administrations are geared towards the efficient handling of day-to-day business – processes and the use of resources are geared towards maximum operational performance. Additional costs are incurred for Strategy Development and Implementation and need to be factored into business planning. We know from countless projects that organisations and administrations can manage Strategy Development with their resources, but stumble when it comes to implementation due to the insufficient availability of knowledge holders. In many cases, they also lack the methodological and technical expertise required to make complex implementation projects a success. We therefore recommend analysing not only the capacity requirements, but also the skills needed and taking measures to reduce the gaps, at the latest during the initialisation of the Strategy Implementation. Partnership building is a key strategic element that should be assessed during Strategy Development. The temporary involvement of external specialists is an alternative. Before implementing the strategy, it is worth taking a look at the strategy elements «Development method« and «Value creation steps«. This allows you to define how you will generally approach Strategy Implementation. Examples are:

- **We do it ourselves**
  - Implementation through the existing line organisation
  - Implementation with a cross-functional, dedicated project organisation
- **We form partnerships**
  - Implementation with existing or new partnerships
  - We do it ourselves first and use partners at a later stage (or vice versa)
- **We are founding a new organisation**
  - Execution outside the traditional organisation (outsourcing to a new company/ startup)

The choice of these strategy elements is based on assumptions that you can test by formulating hypotheses and carrying out tests. In any case, the choice made is an important prerequisite for Strategy Implementation.

The methods to be used for Strategy Implementation require experience in their application. Later in this chapter, we will look in more detail at the question of which methods are useful for which purposes. It makes sense to gain experience with new methods in pilot projects first, rather than in the most important corporate project, that will determine the success or failure of the organisation. If the strategy is implemented without appropriate methodological experience, this is a risk that requires special measures and monitoring or support.

5.5 Methods for Strategy Implementation

Strategy Implementation often has to be coordinated and managed as a whole and is therefore frequently organised as strategic initiatives or programmes. They are based on the concepts of project management and agile development with the methods or frameworks designed for this purpose. Your organisation has probably already successfully implemented many initiatives and employees are familiar with different implementation methods. The aim here is therefore not to impart detailed knowledge of these methods, but to show you which aspects you should take into account during implementation in order to successfully implement your strategy in an agile manner.

In this chapter, we first consider which types of initiatives should be implemented in an agile manner and which should be plan-driven. We then consider whether plans should be implemented through initiatives or processes. You will then learn which aspects are particularly important for agile implementation. Using specific methods and frameworks that are widely used in practice, we will show how methods can be integrated into Strategy Implementation and how impact orientation can be embedded in the methods.

Plan-driven or Agile Implementation?
Depending on the initial situation and the strategic option, different factors determine the success of Strategy Implementation. You are probably already familiar with these from your personal work experience or from the literature on Strategy Development. Nevertheless, a brief reflection is worthwhile because the success factors are closely related to the choice of methods and the management of Strategy Development and Implementation.

If you develop a new product, e.g. an app for analysing data transmitted by pressure sensors, over several iterations, you may be faster than your potential competitors in the market (time-to-market) because you can deliver a first product release with a minimum set of functions (MVP = Minimum Viable Product). You also have the advantage of being able to evolve the product to deliver greater customer value by quickly gathering customer feedback. In such cases, an agile approach is ideal. This can be organised as an agile project or in an agile development process – but it must be agile. We will discuss the project or process further below.

But what if all the requirements for your new product have to be met from day one? For example, a food distribution centre with cold rooms for perishable goods or a new generation of pharmaceutical packaging machines must meet the requirements for function, quality, and safety right from the start. These types of initiatives need to be planned and built as a whole. Implementation is therefore plan-driven. These initiatives are typically organised as a project.

Agile principles can also be applied to the plan-driven approach. For example, elements such as prototypes can be built as needed in order to gain evidence-based insights. Work can also be carried out in self-organised, autonomous teams. It is also possible to develop in stages, each of which leads to a measurable impact. It is also possible to develop parts of the product in an agile manner in a plan-driven project. In these cases, we speak of a hybrid approach.

We can basically distinguish between the plan-driven, agile, and hybrid approaches. The Cynefin framework, which is explained in Chapter 8.1, provides a basis for deciding which approach is suitable for which project.

In the complicated area, we tend to take a plan-driven approach (e.g. construction of a food distribution centre), in the complex area an agile approach (e.g. development of an app). If the initiative includes both complicated and complex elements, a hybrid approach should be considered.

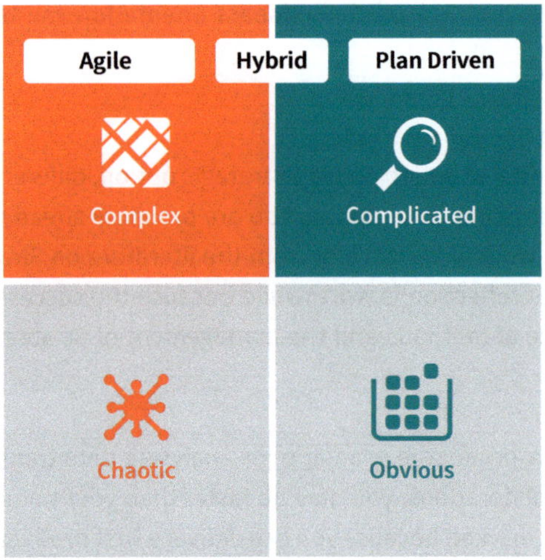

Fig. 5.6: Choice of process method: plan-driven, agile or hybrid

Implementation through a Project or Agile Development Process?

Projects are unique, processes are recurring. When you build a house, it is unique to you in that particular constellation. For the contractor, building your house is also unique, but overall, building a house is a process that is essentially always the same. The construction company is interested in organising the processes in a process-optimised way, so that the highest possible profit can be achieved, because everything runs in a well-established manner. The same applies when you have software developed by a software development company. Basically, it is a question of perspective and the frequency of similar initiatives whether we organise something as a project or as a process. From this, it can be deduced that Strategy Implementation is primarily implemented as a project from the perspective of one's own organisation due to its uniqueness. As mentioned above, this can be organised in an agile or plan-driven manner.

Programme or Project Portfolio?

If several projects need to be carried out to implement the strategy, they can be organised and managed as a programme or as part of a project portfolio. Programmes are limited in time, while project portfolios are managed on a permanent basis and generally include all of an organisation's projects. The decision as to whether the strategy is implemented through individual projects, bundled into a programme or integrated into a project portfolio is determined by the following elements, among others:

- The number and duration of Strategy Implementation projects
- Requirements for coordination and control
- Organisational requirements (existence of a project portfolio)

The programme or portfolio management methods used are applied in addition to Strategy Implementation controlling and do not replace it.

Agile Process Frameworks for Product Development

In specific cases, it is possible to implement product development projects on the basis of process frameworks. This is the case, for example, if you are implementing a large software development initiative that needs to be delivered over a long period of time. The SAFe® or LeSS® process frameworks, for example, can be used to develop and continuously deliver products in defined steps. These methods are primarily used in the IT environment and require a high level of maturity in all disciplines involved, such as business analysis, testing and continuous deployment. Since the process chain is only as strong as its weakest link, all partners must be well-versed in their collaboration. If these frameworks have not yet been integrated into the organisation, this usually requires a concerted agile roll-out between the partners.

> **Practical Tip**
> **Agile Development and Project Management**
> Agile product development frameworks such as Scrum cover the management of the teams, but not all the other aspects of project management such as stakeholder management, risk management, supplier management or the integration of the solution into the organisation. When using these methods for Strategy Implementation, it is important to ensure that the project management aspects are sufficiently addressed. This is why agile development is often embedded in a hybrid project management method.

5

5.6 In-Depth Example: Strategy Implementation with the HERMES Project Management Method

What is the HERMES Method?

The HERMES method is a freely available standard for project and programme management. HERMES is the property of the Swiss Federal Administration and is used not only in administrations but also in many organisations in the private sector. The method is described as simple, understandable, and immediately applicable, can be used free of charge and also includes document templates and checklists. HERMES is available at www.hermes.admin.ch.

HERMES for project management describes the project management process from the client's perspective, from the initial idea to project completion. The backbone of the project management method is a Phase model through which the client controls the progress of the project. At milestones (quality gates), control decisions are made by the client or the project manager. The method includes the description of tasks, outputs and roles and has a modular structure. Depending on the characteristics of the project, certain modules can be omitted. The method can therefore be easily adapted to the specific project content and organisation. HERMES is a hybrid method that combines plan-driven and agile approaches.

The HERMES method can be compared to many other project management methods, but is characterised by the fact that it is results-oriented and quick to apply, because it not only describes the control and management of a project, but also its execution and can be easily adapted to the respective project type. For this purpose, scenarios can be defined, which can be generated online and for which tools such as document templates and checklists are available.

Agile Strategy Implementation and Procedure with HERMES

In order for HERMES (and similar methods) to be used for impact-oriented Strategy Implementation, the following points should be taken into account:

- The project canvas forms the basis for initiating a project. Many organisations using HERMES create a project profile as part of the budgeting process. This corresponds to Strategility's project canvas. The company's project profile may need to be supplemented with information about the impact.
- In the project's initialisation phase, assumptions from the Strategy Development that have not yet been confirmed or refuted are investigated in more detail as required. For example, market analysis may be carried out and solution variants developed. Depending on the information already available from the Strategy Development, the «initialisation« and «concept« phases can be completed very quickly.
- The project assignment (or an implementation assignment) in HERMES is prepared as the basis for the decision to approve the project. The objectives described in the «study« and «project assignment« outputs must be described as impact objectives

(for experienced HERMES users: the system objectives can be subdivided into impact objectives and output objectives, for example). Many HERMES users already do this, but it is not explicitly described in the method.

- At the phase transitions of HERMES, the checklists are supplemented with the impact assessments. The client's control task now also includes the impact assessment.
- Further milestones are to be defined in the course of the project, against which the impact is to be measured.
- The client should be involved in the impact assessment.

The Interaction between the Roles of Strategility and HERMES
The roles in a project or programme implemented with HERMES could look as follows:

Strategility role	HERMES role
Enabler	In the core organisation; decides on project release
Strategy Owner	Project client, programme client
Hypothesis Owner	User representative, business process owner
Strategy Exploration Team	Specialist roles
Implementation Owner	Project and/or programme manager
Implementation Team	Specialist roles

Table 5.1: Relationship between the roles of Strategility and HERMES

5.7 In-Depth Example: Strategy Implementation with the SAFe® Process Framework for Product Development

The Scaled Agile Framework (SAFe®)[9] is widely used in companies and organisations with large IT-based solution and product development projects. Like Strategility, SAFe® is based on agile principles and organises the work into iterations. It is also driven by events with a similar structure and content. The two frameworks, Strategility and SAFe®, can therefore be easily combined.

The considerations and illustrations can also be applied to other popular agile frameworks for large-scale solution and product development such as Large Scale Scrum (LeSS®), Disciplined Agile Delivery (DAD), Kanban Flight Levels or Scrum Nexus. To work with Scrum or (software) Kanban, you need to add your own events, artefacts and roles to your approach. The additions serve the long-term perspective and coordination with other strategically relevant activities and initiatives, such as organisational development.

9 Based on version 5.1 www.scaledagileframework.com.

5

Gain Insights for Strategy Development

Like all agile methods, SAFe® is insight-driven. Experimentation and testing are an integral part of the empirical working methods in agile teams. The testing of assumptions through experimentation and testing can therefore be delegated to agile teams in terms of both method and content. In the continuous supply chain, the work can be found in the first iteration loop of continuous exploration. With the methods of customer-centricity, design thinking and Lean UX, techniques are available for hypothesis-based work (see Chapter 8). Essential-SAFe® describes everything necessary for collaboration in Strategy Development.

At the event level, the «Iteration Review« and «Iteration Planning« events from ScrumXP are linked in SAFe® to the events from Strategy Development (see Chapter 2).

The Strategy Development work is planned in the programme increment planning with the agile teams involved. An epic for strategizing in the programme backlog is used to prioritise strategizing in comparison to all other work and to allocate resources and teams.

Strategy Implementation

Strategy Implementation is based on lean portfolio management. This is included in Portfolio or Full SAFe®. If Essential or Solution SAFe® is used, you create the prerequisites for Strategy Implementation as described in Chapter 5.3.

You **manage** lean portfolio management in SAFe® as follows:

- You use the **strategic topics to ensure** strategic alignment in all development activities. Strategic realignments and course corrections are incorporated into the planning of all work in SAFe® through the strategic topics tool.
- The **portfolio vision** tool is also used to provide long-term orientation for all work within the portfolio.
- The **roadmap** is used to plan the longer-term development and specify the timeline.
- **Milestones** are used to set important strategic deadlines. They also serve to coordinate with initiatives and work that are not managed within the SAFe® portfolio, for example, necessary strategic initiatives with partners.
- **Lean budget** and the guardrails are used to provide the necessary resources for the implementation of the strategy.
- The specific content is commissioned via business and **Enabler** epics in the **portfolio backlog**.

The following instruments are available to monitor Strategy Implementation:

- The **value stream KPIs** are available at the «Portfolio« level to monitor the impact. These are typically tracking and show the success in the long term.
- The **metrics** recommended in SAFe® for organisational development support the control of the accompanying, necessary strategic transformations.

- It is also advisable to include the outputs and insights from the **system demos** and the **Inspect & Adapt workshops** as leading indicators and for continuous learning.

Filling the Roles from Strategility in SAFe®

The roles in Strategility and in SAFe® at the «Team« level are structured in the same way. As part of the hypothesis-based development of the strategy, the Agile Team can take on the role of the Strategy Exploration Team and the Scrum Master acts as the Strategility Master to support the process. Accordingly, the Product Owner can also take on the role of Hypothesis Owner or work together with the Hypothesis Owner.

For implementation projects, the role in SAFe® may vary depending on the scope and scale of the initiative. The following table provides an overview of possible roles for Strategility in SAFe®.

Strategility	**SAFe®**			
Strategility	**Portfolio**	**Major implementation projects**	**Medium implementation projects**	**Strategy Development Small implementation projects**
Enabler	Epic Owner			
Strategy Owner	In the parent organisation; Business Owners			
Hypothesis Owner		Solution Management	Product Management	Product Owner
Strategy Exploration Team	Enterprise Architect	Solution Architect/ Engineering	System Architect/ Engineering	Agile team
Strategility Master		Solution Train Engineer	Release Train Engineer	Scrum Master

Table 5.2: Roles of Strategility in SAFe®

Table 5.2 can also help to find suitable people to form the Strategy Team. Solution or Release Train Engineers are interesting candidates for the role of Strategility Master. Members of system engineering or an agile team are interesting candidates for the Strategy Exploration Team.

Prioritisation and Budgeting of Strategy Initiatives in SAFe®

When the strategic challenge is urgent, strategic initiatives and explorations should be prioritised. As an Enabler, you can prioritise the strategic topics:

1. Essential maintenance – maintaining operational capability
2. Strategic implementation initiatives
3. Operational development – increasing efficiency, quality
4. Business development – improvement of products and services

By allocating resources in lean budgeting and setting the guidelines, you ensure that the strategic initiatives receive the necessary attention. As an Enabler in the role of an epic owner, you can also play a role in managing the portfolio.

Combined Control of Strategility and SAFe®

If speed and agility are important to you for strategizing, then synchronise the Strategility iteration with the programme increment in SAFe®. The Strategility Review & Planning (see Chapter 2.7) is best combined with the Pre-PI Planning Event in SAFe®. This allows you to strategically align the work for the upcoming PI. The strategically relevant system demos, I&A workshops and ScrumXP iterations must be completed and the outputs analysed before the Strategility Review & Planning. Otherwise, insight-based decision-making is not possible.

5.8 In-Depth Example: Strategy Implementation with OKR

Objectives and Key Results (OKR) is a flexible system for agreeing and tracking objectives. OKR as a system or method is available in different forms.

Strategility and OKR share the following principles:

- **Impact orientation:** The objectives describe a desired target state and are formulated in an open-ended manner. The objectives correspond to tracking indicators.
- **Measurable outputs:** The key results describe concrete, measurable results for achieving the objectives. The key results correspond to the leading indicators.
- **Cascading:** OKRs are defined for long, medium and short-term periods and are related to each other hierarchically. Common terms are:
  - Long term – three to five years: Strategic or long-term Goal – the GOAL
  - Mid-term – one to three years: Tactical or medium-term Goal – the MOAL
  - Short term – one quarter: the OKRs
- **Participation:** The stakeholders formulate the OKRs for their area of responsibility themselves. Employees and teams formulate the short-term OKRs based on the medium-term goals.
- **Rhythm and adaptation:** The objectives are checked against the overarching objectives after each quarter and reformulated for the following quarter.
  Key results are reviewed, corrected or reformulated on a weekly or bi-weekly basis or brought back on track through corrective measures.

Thanks to their impact orientation, rhythm and adaptation, OKRs are very well suited to cognition-driven processes, i. e. more complex processes (see also Chapter 5.5).

These similarities make OKRs very suitable for managing Strategy Implementation. Are you already using OKRs in your organisation? Then the use of OKR is also recommended for Strategy Orientation and Development.

OKR in the Strategy Orientation

In the Strategy Orientation, you set long-term and far-reaching goals. OKRs allow you to communicate these in a tangible and comprehensible way. When the objectives are formulated to guide action, you create autonomy in the workforce. Autonomy promotes agility.

Overcoming the **strategic challenge** is the objective, the desired impact. Tangible key results are the assumptions of uncertainty and the visible signs of change. These key results need to be removed or reduced. The expected effects of a new strategic direction are also key results that should emerge over time.

It is advisable to refine the strategic option using OKRs. The objective describes the desired impact of the strategic option, i.e. the overriding intention being pursued with the strategic option. Key elements of the strategy are suitable as key impacts. Every strategy contains a few elements that determine the success or failure of the strategy. These are your key impacts in the strategic option.

Managing Strategizing with OKR

Both the OKR method and the Strategility framework plan in the medium term with a regular cycle. Considerations on the length of the cycle and the description of the control in Strategility can be found in Chapter 2.7. It is therefore best to combine the «Review & Planning« event with the review and re-planning of the OKRs.

Strategy Implementation is managed via MOALs. These are formulated by the Strategy Team. For implementation, the MOALs in planning are broken down into OKRs by individual teams in the organisation and, if necessary, distributed over several iterations.

To achieve the best strategic alignment in the organisation, you should ensure transparency across all MOALs and OKRs. To do this, you manage all long-term goals, MOALs and OKRs via the Strategility Board. The Strategility Master also masters the OKRs and facilitates the Review & Planning workshops. The role of the Enabler is played by the management.

You may be wondering how many MOALs you can assign? Probably far fewer than you need. Working on strategic MOALs competes with day-to-day business. It is work on the system. Individuals and teams should therefore ideally only pursue one strategic MOAL.

CHAPTER 6

The Strategy Team and its Roles

6 The Strategy Team and its Roles

6.1 Successful Collaboration is no Coincidence

When you initiate the strategizing process in your organisation, you will be thinking about whom you want or need to involve. Other important questions are whether and how Strategy Development can be carried out with autonomous teams, how alignment between those involved can be achieved, who makes which decisions, and what the responsibilities and competencies of those involved are. The Strategility role concept answers many of these questions.

In this chapter, you will learn what the roles are in the agile strategy process and how they can be organised. For the application, we go into more detail on how the roles are filled depending on the actual functions in your organisation. It's like in theatre and film: the script determines the plot, but the actors add the real spice to the play. The role descriptions and behavioural observations of the actors will help you when filling roles and moderating strategy workshops. You will receive tools that you can tailor to the needs of your organisation.

Obviously, the existing and desired future corporate culture must also be taken into account when organising the agile strategy process. We will go into this in more detail in Chapter 7, but we will look at some role-specific cultural aspects below.

Once the issues surrounding the organisation of the strategy process have been clarified, you have a very good chance that the collaboration in agile Strategy Development will not only be energetic and successful but will also be enjoyable and create a shared sense of achievement.

6.2 The Five Roles in Agile Strategizing

The Strategy Team consists of five roles: Enabler, Strategy Owner, Hypothesis Owner, Strategy Exploration Team, and Strategility Master. The roles are filled by employees who already have one or more functions in the organisation. Figure 6.1 shows some typical roles.

Different methods with their own roles are used for Strategy Implementation, for example, the roles of project management or agile development. For the sake of simplicity, we use the generic roles of «Implementation Manager« and «Implementation Team« in the chapter on Strategy Implementation. These will not be discussed further in this chapter.

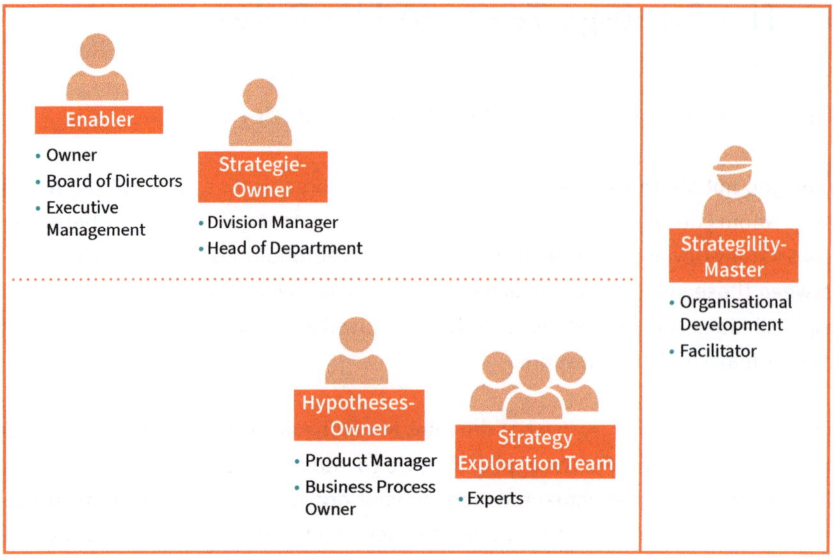

Fig. 6.1: The five roles of agile strategizing

6.2.1 Enabler and Strategy Owner

The Enabler provides the boundaries for agile strategizing. They set the strategic option and, together with the Strategy Owner, define the Strategy Orientation parameters (North Star with mission, vision/guiding principles and values). The enabler is responsible for the company's success and is hierarchically positioned at the highest management level of the organisation.

Enablers and Strategy Owners ensure that the resources are deployed to deliver optimal value to the organisation. Resources include human, financial, production, knowledge, and information resources, as well as anything else of value that the organisation owns, such as relationships with business partners, long-term customer loyalty, etc.

Enablers and Strategy Owners focus on the impact of the strategic option and manage the use of resources with regard to the impact to be achieved. The primary objective of this impact management is to ensure the long-term sustainability of the organisation and the achievement of the overall business objectives.

Enabler and Strategy Owner
Enablers and Strategy Owners serve the organisation by taking the overall strategic view and ensuring that the consequences of decisions are sustainable and future-proof for the organisation.

Who is an Enabler?

In principle, the question of who fills the role of the Enabler is the same whether the strategy is developed using an agile or a more traditional linear approach. However, it is largely determined by the organisation in which the strategy is developed. In the following, we look at the role in the context of three different types of organisations.

1. **The owner-managed company**
   In an owner-managed company, the owner is personally responsible for the overall success of the business and the financial, legal, and personnel consequences of the business activities and Strategy Development. Accordingly, the owner personally assumes the role of Enabler.
2. **Public limited company**
   In the case of a listed company, the ultimate decision-making authority to make decisions about the use of resources rests with the board of directors. For a few strategic decisions, such as mergers of companies, shareholders can intervene by virtue of their shareholding, but strategic issues are not usually decided by shareholders at general meetings. Shareholders therefore do not play the role of Enablers, but the board of directors and/or management do.
3. **Public administration/authority**
   In public administration, the ultimate Enabler is the highest state body, which in a democratic state is usually the parliament or, in a municipality, the municipal assembly, which decides on laws and budgets and can therefore influence the strategy. However, the parliament delegates Strategy Development to the administration. The role of the Enabler is then performed by the relevant top administrative hierarchy. Depending on the organisation and the importance of an issue, the role is delegated to the administrative unit responsible for the issue, for example, to an office for environmental protection, which develops environmental protection strategies and has them approved. In this case, the head of the office or a head of department takes on the role of Enabler.

Who is the Strategy Owner?

The role of Strategy Owner is filled by the person who has overall responsibility for the services and products concerned. This is often the function of a division or department head. The person who takes on this role should be at top management level, e.g., a member of the executive board.

6.2.2 Hypothesis Owner and Strategy Exploration Team

The Hypothesis Owner and the Strategy Exploration Team refine the strategy elements. Based on the chosen strategic option, they define the hypotheses and tests, conduct, and analyse them. Thanks to the empirical approach, they generate a continuous increase in knowledge with each iteration. As the level of maturity of the Strategy Development increases, they confirm or refute the prospects of success of a strategic option.

The Hypothesis Owner and the Strategy Exploration Team are responsible for the quality and significance of the information obtained.

> **Hypothesis Owner and Strategy Exploration Team**
> The Hypothesis Owner and the Strategy Exploration Team serve the organisation by defining and conducting open-minded tests and openly sharing insights on strategy elements and strategic options with Strategy Owners and Enablers.

Who is the Hypothesis Owner?
The role of the Hypothesis Owner is typically played by a product manager, a service manager, or a business process manager.

Who is in the Strategy Exploration Team?
The role of the Strategy Exploration Team is filled by the required experts and knowledge holders. The composition may vary depending on the test.

6.2.3 Strategility Master

The Strategility Master supports and advises everyone involved in the Strategility process. They define the Strategility process together with the participants, facilitating and adapting the strategy process. They also take responsibility for the process.

> **Strategility Master**
> The Strategility Master serves the organisation by shaping the working culture in accordance with agile principles, empowering the participants, moderating with an open mind, and supporting continuous process optimisation.

Who is a Strategility Master?
The role of the Strategility Master is filled by an organisational developer, a strategy manager, a facilitator, or a consultant.

> **Practical Tip**
> **Process or Content?**
> When filling the role of Strategility Master, a key question is whether the person is limited to moderating the strategy process or whether they also work on content-related topics and, for example, prepare analyses or participate in the development and evaluation of strategic directions.
> The role of the Strategility Master is entirely focused on designing and moderating the process. The other roles are responsible for the development of

outputs. However, an important task of the Strategility Master is to check the consistency of the outputs and to moderate the process in such a way that consistency is ensured and contradictions are identified. They also provide support and advice in the definition of hypotheses and tests.

6.3 Considerations on the Formation of the Strategy Team

Who Should be Involved in Strategy Development?
If you are responsible for Strategy Development, you want to achieve optimum results, but you also want the process to be quick and efficient. In addition, the strategy should be supported in such a way that it has the necessary acceptance for Strategy Implementation. It is therefore important to fill the roles with the «right« people. There are a number of key questions that can be answered when putting together the initial Strategy Team.

> **Practical Tip**
> **Key Questions for Forming the Strategy Team**
> The following considerations can help you decide who should be part of the Strategy Team:
> - Who should be involved based on their role in the organisation?
> - Who should be involved on the basis of their knowledge or special skills?
> - Who should be involved based on their formal or informal position in the organisation?
> - Who outside the organisation should be involved in Strategy Development, and what role would they play?
> - Who should be involved to ensure acceptance of the strategy for Strategy Implementation?

Can or Should One Person Play Several Roles?
The following experiences can be helpful when filling roles.

Role cumulation can have a reinforcing effect

In principle, one person can take on several roles (role cumulation). Role cumulation can strengthen autonomy and alignment, because coordination is easier when one person wears «several hats« in terms of different functions or roles. For example, a person with the role of Enabler may also be the Strategy Owner. This is often the case in owner-managed companies.

The Hypothesis Owner may be a member of the Strategy Exploration Team in the role of a tester.

It is promising if the future person responsible for an implementation initiative already fulfils the role of Hypothesis Owner.

6

Role cumulation can have an inhibiting effect

For alignment to be easy, the Strategy Owner and the Hypothesis Owner need to be in close dialogue. If the same person fulfils both roles, the exchange is undoubtedly immediate. However, there is a risk that a bias could lead to a lack of open-minded verification of the defined hypotheses, meaning that tests and analyses are consciously or unconsciously biased towards confirming the hypotheses and assumptions or that falsification (refutation) is undesirable.

In one company, for example, we found that the Strategy Owner, who was also the Hypothesis Owner, ignored the test results even though they clearly showed that the tested product was not popular with the target customer group.

The Strategility Master should not take on the additional roles of Enabler, Strategy Owner, and Hypothesis Owner, as this usually makes open facilitation more difficult.

Organisation of Several Strategy Teams Working in Parallel

If several strategy elements are to be explored and developed at the same time, it can be time-saving to set up several teams. The work of the teams then needs to be coordinated. This can be supported as follows:

- Enablers and Strategy Owners synchronise the content of the work. This typically takes place with an appropriate iterative synchronisation event, such as the Review & Planning meetings.
- The Strategility Master manages the coordination activities in the Strategility backlog and manages the corresponding synchronisation event, which has to be defined specifically for this purpose.
- Hypothesis Owners and Strategy Exploration Teams regularly exchange insights.

In order for the teams to be able to work on Strategy Development in parallel, the North Star with a clear definition of the mission/vision and values is required to support alignment. However, the central element is the defined and published strategic option to which the teams are aligned.

6.4 Roles and Events in the Strategy Process

Events as Elements of the Control System

The roles of Strategy Owner and Hypothesis Owner are not only responsible for outputs and impact, but also directly control events as participants by making decisions. Figure 6.2 provides an overview of the events.

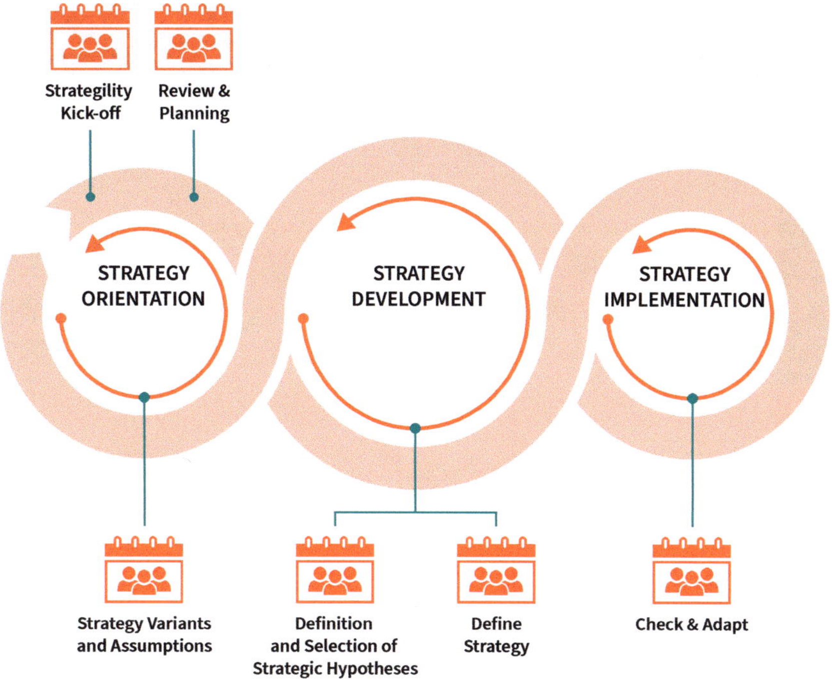

Fig. 6.2: The events in the process flow

Strategility Board for Communication, Control and Coordination

The Strategility Board is the central tool for facilitating communication and controlling and coordinating the work. It is divided into one horizontal lane for each of the three sub-processes (Figure 6.3) and the backlog is prioritised by the respective managers.

- The alignment path is the responsibility of the Enabler and Strategy Owner.
- The development path is the responsibility of the Strategy Owner, Hypothesis Owner and Strategy Exploration Team.
- The implementation path is the responsibility of the Implementation Owner.

The board is managed in the «Review & Planning« event, among others. Figure 6.3 shows the Strategility Board with the alignment, development, and implementation paths.

Read more about the control system in Chapter 2.7.

Responsibility in the Process Steps

Efficient and effective Strategy Development requires that it is clear who is involved in which process steps, who takes responsibility for which process steps and how decisions are made based on the insights gained.

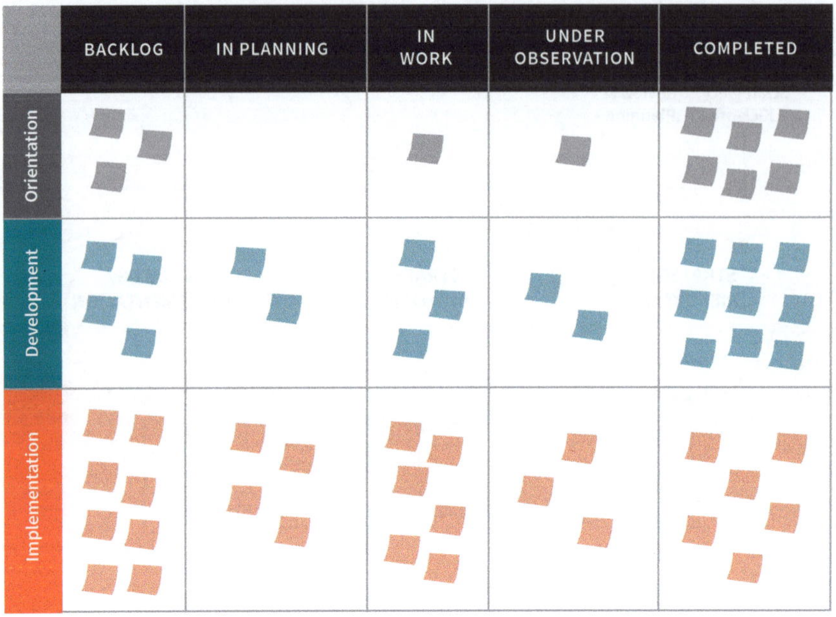

Fig. 6.3: Agile management and coordination of work in the three areas of the strategy backlog

Figure 6.4 shows the implementation responsibility and involvement of the roles along the tasks in the strategy process. The minimum number of participants is listed. Depending on how the strategy process is opened up, additional knowledge holders will be involved in the tasks (see Chapter 1.6).

Chapter 8.7 deals in detail with how knowledge is gained and decisions are made. It lists who makes which decisions. This is related to the responsibility of the roles for the process steps shown in Figure 6.4.

> **Practical Tip**
> **Several Strategility Masters**
> In organisations with several business units, it has proven effective to have several people in the role of Strategility Master who support the board members in their business unit. As a team, they define and anchor the Strategility process together.

Distinguishing Responsibility for the Impact from Responsibility for the Outputs

Particularly in organisations with a pronounced agile mindset or in organisations with a strong focus on values, we have found in the course of our consulting work that there is often a need to clarify who makes which decisions. Distinguishing between responsi-

A Accountable R Responsible

PROCESS STEPS / ROLES	Enabler	Strategy owner	Hypothesis owner	Explorations team
Identify strategic challenge	A	R		
Adapt / derive mission statement (mission, values, vision)	A	R		
Determine strategic direction and review on an ongoing basis	A	R		
Define strategy elements		A	R	
Determine most impactful assumptions and define hypotheses		R	A	R
Define and carry out tests			R	A
Define and continuously review strategy		A	R	
Adopt strategy and release for implementation	A			
Implement strategy		A		
Strategy implementation impact and result evaluation	A	R	R	

Fig. 6.4: Tasks and roles with the allocation of responsibility and participation

bility for the impacts and responsibility for the outputs helps to clarify the underlying issues and sharpen the understanding of roles.

Figure 6.5 shows that people who take on the role of Enabler or Strategy Owner are responsible for the impact of the strategy – i. e., the consequences of success and failure. The people who take on the role of Hypothesis Owner or Strategy Exploration Team are responsible for ensuring that the outputs they produce are correct and meet the requirements set.

It is therefore conceivable that the defined requirements for an output may be met, but the impact cannot be achieved and the strategy has to be adjusted. It can be concluded that the Hypothesis Owner and the Strategy Exploration Team make the decisions that are related to the development of the output. The Enablers and Strategy Owners, on the other hand, make the high-impact decisions, such as defining the strategic option and strategy. The case study illustrates this clearly. Chapter 5, on Strategy Implementation, explains the topic in more detail.

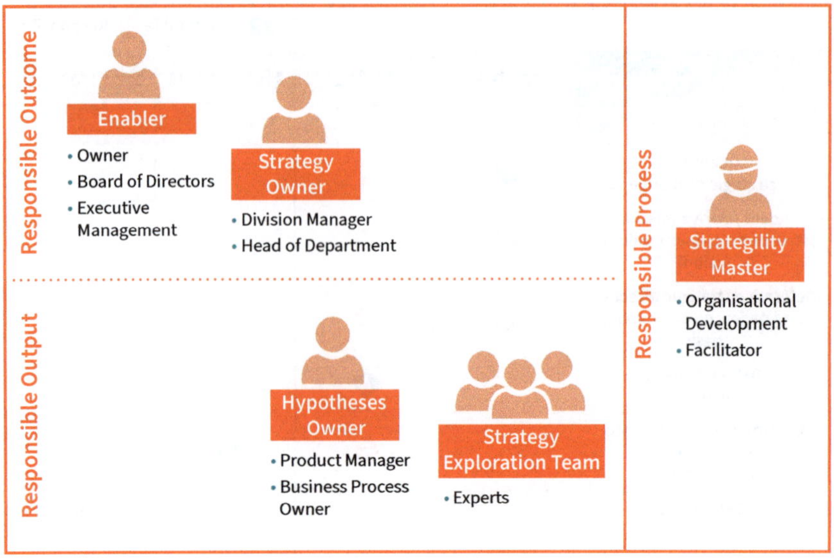

Fig. 6.5: Roles and their responsibility for impact, results and process

> **Example**
>
> **Good Project Implementation Does Not Guarantee Strategy Success**
>
> The Swiss airline Swissair fell victim to its own expansion at the beginning of the millennium. It pursued a growth strategy by acquiring other companies and acquired several airlines in a short period of time. The integration of the acquired airlines into the parent company took place in record time through professionally managed projects. The technical integration projects were very successful, achieving their objectives on time and within budget. However, the airline ran into difficulties, partly due to the poor financial situation of the acquired companies, which ultimately led to the grounding and bankruptcy of Swissair. The outputs of the Strategy Implementation were technically successful, but the impact of the strategy was disastrous. The company's demise cost several thousand jobs. The responsible Enablers were subsequently taken to court to clarify responsibilities.

In the *Digital Extras* you will find a canvas to visualise your Strategy Team.

6.5 Understanding Roles and Conflicts

In Strategy Development, we go through different stages of abstraction and concretisation. Collaboration within the Strategy Team is usually strongly influenced by the personalities involved. Depending on the mindset, interests, character, or claim to authority and power, conflict situations arise that hinder the work process or diminish the output. We also observe different ways of thinking and behaving, i.e. personality types, which have a strong influence on the process.

Doer or Expert

The roles of Enabler and Strategy Owner described above can be filled by different personality types.

In the agile Strategy Development workshops, we found that some of the Strategy Owners are a mixture of «innovators and doers« (Belbin 2004). They are often the drivers of strategic initiatives and develop their ideas in collaboration with interested, well-known, and well-meaning customers and business partners. The resulting positive energy and resonance encourages them and serves as proof of their idea. They are often convinced of their initiative and find the structured derivation of the assumptions behind their idea tedious and a waste of time. Further structured testing of the idea seems like unnecessary effort to them – as they have already discussed it with their interlocutors. As a result, they quickly become impatient in our Strategility workshops and are reluctant to continue their thought processes.

Another group of Strategy Owners reminded us more of the personality types of evaluators. They are convinced co-thinkers when it comes to explicitly formulating the assumptions behind a strategic idea and the conceptualisation of test possibilities for these assumptions. They are intrinsically motivated to evaluate the strategic importance of a proposed idea and to analyse its meaningfulness and feasibility.

Both types have their importance as Hypothesis Owners in agile Strategy Development, but in strategy workshops they are often in contrast with each other in terms of their interests, their way of thinking, and their pace of decision-making.

Practical Tip

Typical Conflict Situations You Should Prepare for

You should expect these situations:

- How do you react as an Enabler when you receive feedback from the test team that shows that the idea is not promising?
- How do you react as a Strategility Master when you realise that the test team is not open to results and is only looking for confirmation of the hypotheses?
- As a member of the test team, how do you react when you realise that the Enabler is not taking the insights from the tests into account in their decisions?

Here are some suggested actions:

- Conduct a role clarification at the kick-off event to address and clarify the different personality types, their significance in the process steps, and the mutual expectations.
- To discuss the test results, organise an appropriate event that specifies how to handle them.
- Define quality criteria and checks for test definitions and test these in appropriate ceremonies.

6.6 Role Descriptions for the Practice

The role descriptions, which you can also find in the *Digital Extras*, can be customised for use in practice. The roles with their tasks, responsibilities, and competencies are described below.

Enabler
Tasks:
- Defines the strategic benchmarks and defines the strategic options
- Ensures the optimal use of resources across the organisation
- Creates the conditions for successful Strategy Implementation
- Empowers the teams by allocating the necessary resources (people, finances, etc.)
- Defines the members of the Strategy Team
- Reviews the impact of the strategy and makes adjustments to the strategic option if necessary

Expertise:
- Prioritises the initiatives based on resource allocation
- Decides on Strategy Orientation parameters
- Decides on strategic options
- Decides on the strategy option to be pursued
- Decides on the organisation and role holders in the Strategy Team
- Decides on the approval of the Strategy Implementation

Responsibility:
- Responsible for the overall success of the company and the impact of the strategy
- Effective use of resources
- Supports the cultural change required for agile strategizing and acts as a role model

Required skills:
- Understands the principles of agile Strategy Development and agile working methods and has an agile working mindset
- Distinguishes between responsibility for impact and output and focuses on the impact in their work

Strategy Owner

Tasks:

- Ensures the development of the strategy with the strategy elements for the defined strategic option
- Defines the assumptions underlying the strategic option, and the strategy elements
- Takes the feedback of the insights from the agile Strategy Implementation, analyses the impact on the strategy elements and the strategic option and makes adjustments where necessary
- Participates in the defined events
- Advises Enablers and Hypothesis Owners
- Leads the Strategility Board in their area of responsibility

Expertise:

- Decides on the strategy with the strategy elements (as a prerequisite for the decision to release the Strategy Implementation by the Enabler)
- Decides on measures to empower the Strategy Team

Responsibility:

- Responsible for the impact of the strategy elements
- Effective use of resources for Strategy Implementation
- Supports the cultural change required for agile strategizing and acts as a role model

Required skills:

- Understands the principles of agile Strategy Development and agile working methods and has an agile working mindset
- Distinguishes between responsibility for impact and output and focuses on the impact of the strategy in their work

Hypothesis Owner

Tasks:

- Participates in the defined events
- Develops and analyses the assumptions underlying the strategic option and the strategy elements and derives hypotheses
- Defines the tests together with the Strategy Exploration Team
- Analyses the evaluation of the tests and determines the next steps
- Discusses the test results with the Strategy Owner
- Advises the Strategy Owner on alignment with strategy elements and strategic option
- Leads the Strategility Board in their area of responsibility

6

Expertise:
- Decides on tests and their prioritisation with the Exploration Team

Responsibility:
- Open-minded analysis of assumptions and definition of the hypotheses
- Open-minded assessment of the test results
- Quality of the outputs

Required skills:
- Understands the principles of agile Strategy Development and agile working methods with an agile working mindset
- Distinguishes between the responsibility for the impact and the output and focuses in their work on the outputs and their impact on the impact of the strategy elements

Strategy Exploration Team
Tasks:
- Analyses the assumptions and hypotheses
- Defines the tests
- Runs the tests, analyses them, and documents the outputs
- Validates tests and makes optimisations
- Explains the insights to the Hypothesis Owner
- Leads the Strategility Board in their area of responsibility

Expertise:
- Decides on tests and their prioritisation with the Hypothesis Owner

Responsibility:
- Open definition of the tests in terms of value and results
- Performance of the tests and quality of the outputs
- Open-minded evaluation of the test results

Required skills:
- Understands principles of agile Strategy Development and agile working methods with an agile working mindset
- Experience in test design, verification, and optimisation
- Experience in test execution and evaluation

Strategility Master
Tasks:
- Supports the application of the Strategility process
- Acts as a moderator and facilitator
- Manages the implementation of events
- Ensures continuous process improvement

- Reviews the outputs of the Strategy Development for consistency
- Carries out impact, output and process reviews during Strategy Implementation and ensures feedback from Strategy Implementation to Strategy Development
- Leads the Strategility Board with process elements

Expertise:
- Moderates the events
- Triggers process optimisation
- Has access to all information and outputs to ensure consistency of outputs

Responsibility:
- Process management for the entire Strategility process
- Open-minded moderator
- Ensures that those involved in the process fulfil their role

Required skills:
- Experience in strategy facilitation and strategy documentation
- Experience with agile working methods and an agile mindset
- Can communicate the principles of agile Strategy Development and agile working methods to those involved in the process and to stakeholders in the environment
- Experience in test design, verification and optimisation
- Experience in Strategy Implementation through plan-driven and agile initiatives
- Experience in project management with impact measurement, output review and process review
- Experience in test execution and evaluation

CHAPTER 7

Strategility Culture

7 Strategility Culture

In the previous chapters, you have been introduced to the framework for agile Strategy Development. You may have already realised at certain points that some of it fits your organisation perfectly, while others may be more difficult. The reason for this is usually the corporate culture. At this point, the question arises: What kind of corporate culture is important to be successful in Strategility? Or to put it another way: Does Strategility work in all cultures? To answer this question: Yes, you can use Strategility in all corporate cultures if you take the cultural aspects into account. In the following, we explain the background to the Strategility culture and make recommendations for shaping the culture.

7.1 Operating Principles for Strategility

Strategility is a combination of two concepts with different operating principles: the concept of strategic management and that of agility (see Figure 7.1).

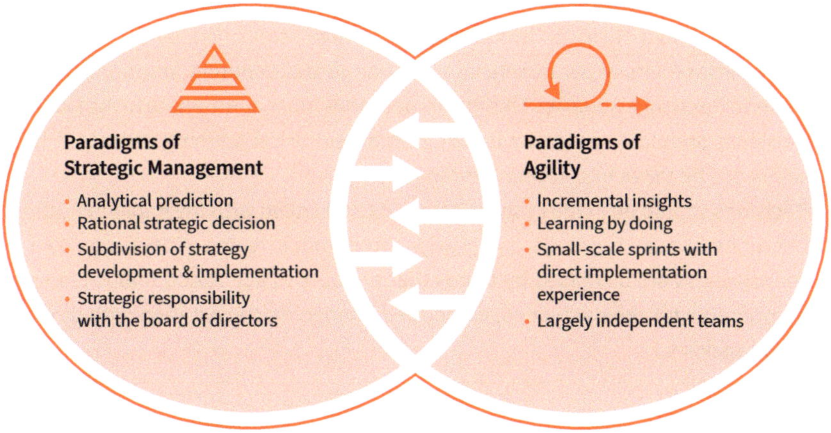

Fig. 7.1: Strategility paradigms from two concepts

Strategic management seeks to define and implement long-term, sustainable goals. It aims to make the assumption of entrepreneurial risk assessable and successful. Strategic management assigns responsibilities and tasks for the development of the company. The following principles have therefore become established in strategic management:

- Decisions should be based on analytical forecasts. Detailed analytical methods for assessing external and internal company conditions are standard in any strategy process. Tools such as the PESTEL environment analysis or the 5 Forces industry

analysis are used to ensure that long-term strategic decisions take future developments into account as accurately as possible.
- Decisions should be made rationally. They therefore follow a rationality that is as comprehensible as possible and can be understood and assessed by business partners, employees, and investors.
- Decisions should be separated from their actual implementation. On the one hand, this makes it possible to share responsibilities for both correct decision-making and correct implementation. On the other hand, it also allows different skills and interests to be brought to bear. Not every employee wants to learn about complex decision-making issues and take responsibility for their decisions, and not every manager wants to acquire and apply the detailed technical expertise that is required in many positions on a daily basis.
- Leaders should be accountable for decisions. The ability and willingness to deal with complex development opportunities and to take responsibility for the corresponding decisions is concentrated in the management functions of an organisation. This allows financial and legal consequences to be specifically allocated.

The concept of **agility** strives for the long-term existence of an organisation through adaptability. The organisation anticipates change and takes the initiative to respond to it. The following principles guide the organisation's actions:
- Decisions should be driven by iterative knowledge acquisition. Small change steps with quick results allow beneficial activities to be identified and non-beneficial paths to be changed quickly. This makes decisions more appropriate to the situation.
- Decisions should be made by learning. In an empirical learning process, new insights can be taken into account iteratively.
- Decisions should be made in short iterations and linked with the implementation.
- Decisions should be made by largely autonomous teams. Knowledge and responsiveness are increased by teams that are able to act. Autonomous teams are accountable for their activities, embedded in ongoing consultation with other decision-makers.

7.2 Analysing Corporate Culture for Strategility

To implement both concepts, supporting attitudes and behaviours are required – they are considered an expression of the cultural essence of an organisation. Cultural analysis according to William Schneider (Schneider 1994) has established itself as a means of assessing the cultural prerequisites for agile corporate development. He distinguishes between four cultures along the two axes (see Figure 7.2):
- horizontal: person-oriented vs. company-oriented
- vertical: reality-oriented vs. possibility-oriented

Fig. 7.2: Types of culture based on Schneider (1994)

Take a moment to categorise your organisation. You can do this either ad hoc, using the illustrative terms in the four cultural quadrants, or with the help of the more detailed self-assessment for cultural analysis in the *Digital Extras* of the book. Below is an extract from this.

Extract from the Cultural Analysis according to Schneider (1994)

2.1 At the end of the day, we will only be successful in this organisation if we:

- Create the conditions for people within the organisation to develop and deliver valuable performance.
- Create an organisation that has a high level of expertise and makes use of it.
- Put together a group of people who form a team and the team members use each other as a resource.
- Have and maintain control.

2.2 What do we primarily pay attention to in our organisation and how do we make decisions?

- We pay attention to what is, and we make decisions by listening to our heart and our head.
- We pay attention to what could be, and the basis for our decision is an objective and detached analysis.
- We pay attention to what could be, and we rely on what develops in the hearts and minds of our employees when making decisions.
- We pay attention to what is, and base our decisions on an objective and detached analysis.

Depending on the results of your cultural analysis, this will have different implications for your Strategility work. In the following, we explain the significance of your Strategility culture and the options for shaping it.

Practical Tip

Strategility Meaning of the Culture Types

- If you have categorised your company as belonging to the people-oriented culture types of **collaboration** and **cultivation**, you should have a good basis for the agile development of your organisation and service offerings.
- If you have assigned your company as belonging to the company-oriented culture types of **control** and **competence**, you should have a good basis for developing your organisation and service offerings in line with strategic management.
- If you have classified your company **between the people-oriented and company-oriented culture types**, you should have a good basis for developing your organisation and services in line with the Strategility approach.

What should you do if you find that your cultural type is a challenge for Strategility?

- Do you have a strong cultural type and the counterpart for Strategility is less pronounced?

Example

Gap in the Strategility Culture

A manufacturer of pressure measurement sensors had been growing successfully for 50 years. The development of products and customer groups was the result of opportunities and ideas. The Strategy Development process and the underlying corporate culture had agile characteristics without being consciously designed: Strategy changes – such as new products and customer segments – were developed collaboratively, new products were iteratively brought to market, and customer feedback was gathered and processed based on evidence.

Increasing competition, complex strategic issues such as the expansion into India, and a high operational workload led to the realisation that more strategic planning, explicit task allocation, expertise and clear orientation were needed in some areas. These missing behaviours were the gap for an agile and strategic corporate culture.

- Do you have strong characteristics at opposite poles that lead to tension?

> **Example**
> **Tensions in the Strategility Culture**
> At the pressure sensor manufacturer, the favourable cultural aspects for agility and those for strategic management are distributed differently within the company. One member of the management team, along with engineers and developers, tends to adopt behaviours that are more conducive to agility. The other member of the executive board and functional managers tend to adopt behaviours more in line with strategic management. The two board members have become aware of this difference and have been coached together in order to develop a way of dealing with this situation that adds value.

7.3 Recommendations for Shaping the Strategility Culture

Changing a corporate culture completely takes years and ties up a lot of energy in the company. It is therefore advisable to work with the existing culture rather than against it. The following activities can directly promote the ability to act strategically:

- Creating organisational units that are agile and take action, and those that are strategically managed and take action. In organisational development, we speak of two operating systems: you align departments and strategy discussions according to strategic management, where plannable, linear decision content is given. Departments such as accounting or purchasing and strategic issues such as the continuation or optimisation of an established market position are often mentioned in this context. At the same time, create agile organisational units that act to identify potential threats and opportunities and use them to implement creative strategies at high speed. In the current debate, this is referred to as your second, agile operating system. Iterative, change-intensive issues are dealt with in the relevant departments and working groups.

> **Example**
> **Two Operating Systems**
> The manufacturer of pressure sensors has set up ad-hoc working groups over the years. Employees and managers from different departments come together on their own initiative when they have ideas for a change initiative and their expertise is needed to shape the change. These ad hoc working groups work and decide autonomously within a certain financial and change framework. Past topics of such ad-hoc working groups have included the introduction of semi-autonomous production units, a possible expansion into Asia, or entry into the hydraulics market. They disband when the issue is decided or is not ready for a decision.

At the same time, a functional organisational structure exists in parallel with the often agile ad hoc groups, within which work processes are planned and partly standardised.

- Combine representatives of agility with representatives of strategic management in a development process. Development processes include agile steps such as idea generation and iterative knowledge acquisition as well as strategic steps such as planning and reflection. Make the different behaviours of agility and strategic management explicit and clarify their added value for a development process. Ask representatives of these behaviours to be clear about their importance in the respective process steps, especially if they take on explicit strategic roles. For example, a Strategy Owner is more likely to have a strategic management background and mindset. They should be aware that such behaviours add value to strategy and alignment, especially when evaluating insights and feedback. When developing hypotheses or tests, strategic management behaviours can be more of a hindrance, as they require learning and adaptation rather than knowledge and planning. This may mean recognising that the Strategy Owner is not the focus of these process steps and should hold back somewhat. The Review & Planning events in particular can be a crystallisation opportunity for the conscious management of the two basic attitudes and should therefore be designed and moderated by the Strategility Master with this in awareness.

Practical Tip
Method to Support Mutual Appreciation
To create awareness and appreciation of different attitudes, you can use the following facilitation method. You will need approx. 20 minutes in small groups and then approx. 10 minutes presentation time per small group. Ask the small groups to draw «the typical agilist« or «the typical strategist« on a flipchart. A sketch in the form of a stick figure is sufficient. The small group enriches the stick figure with characteristic features, such as a clock to symbolise an iterative approach or a large head to symbolise constant thinking. The aim is to characterise these typical representatives with a loving wink.
After sketching, each small group completes the following sentences:
The typical representative of agilists/strategic management ...

- is characterised by ...
- is notorious for ...
- is appreciated for ...

We would like to express our gratitude for ...
This method can be used, for example, in the «Strategility Kick-off« event.

- Consciously use methods that support the behaviour you want to promote: For the agile learning culture for example, define tests that can be carried out and analysed within a day, including the question: What do we learn if the test contradicts our assumptions (see Chapter 8)? For the systematic planning and reflection of tests, use e.g. the test canvas (see Chapter 4.4).

All in all, you can use Strategility in any corporate culture – with conscious support for the cultural aspects you want to promote. Emphasise the different aspects according to your development needs.

For more agility, a transition is needed:
- from result to impact orientation
- from «knowing/planning« to «learning/adapting«
- from commanding to steering
- from isolated management decisions to shared development

To be strategic, management requires a transition:
- from action to reflection
- from promotion to rational explanation
- from being operationally busy to taking time to plan
- from helping out to taking responsibility

CHAPTER 8

Hypothesis-Based Decision-Making

8 Hypothesis-Based Decision-Making

8.1 Deciding Wisely in the Face of Uncertainty

The more ambiguous the situation, the more uncertain a decision is, the more important your role as a manager becomes. Uncertainties can have different causes, and the context, the environment in which decisions are made, can be different. Different decision-making processes are therefore required in order to make intelligent decisions in the face of uncertainty.

Based on the Cynefin framework (Snowden 2021) in Figure 8.1, complex and complicated situations are relevant for strategic decisions. From a management perspective, decision-making in chaotic situations is more likely to be associated with crisis management. Decision-making under clarity is mainly found in operational activities.

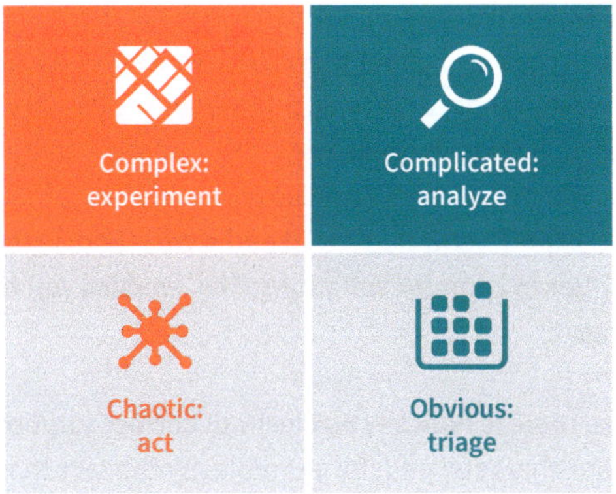

Fig. 8.1: Decision-making under uncertainty in the Cynefin framework according to David John Snowden

The terms «complex« and «complicated« are often used synonymously. In the VUCA context and with the Cynefin framework, the situations can be distinguished as follows:

Complicated Situations: Analysis
If a question can be solved or decided analytically – with specialist knowledge and expertise – it is a so-called complicated situation. In complicated situations, there are models or scientific laws of cause and effect that explain the relationships. Analysis and careful clarifications can fill in information gaps and eliminate uncertainties. This enables you to make informed decisions.

In complicated situations, there are usually several correct solutions, and the majority of experts agree. Traditional Strategy Development in times of stability relies heavily on analytical methods for complicated issues.

Typical Analysis Examples

In order to determine the likelihood of the «market development« strategic option, it is essential to assess the market potential. For goods and products with many customers, market data can be obtained from official statistics, industry associations, or specialist consultancies. This applies, for example, to the online newspaper in Chapter 3 and to the retailer/rental company in Chapters 1 and 2.

Assessments of logistics, construction, and infrastructure are strategic issues that are solved conceptually/analytically.

When developing technical solutions for devices, machines, or even software, the creation of variations of existing products and solutions can often be carried out largely analytically. In project management, this procedure is also known as the Big Design Upfront (BDUF). This is typically the case with the strategic option of «market penetration and development«.

Complex Situations: Experimenting

In complex situations, cause and effect can only be determined retrospectively. Complex situations are characterised by the interdependence of different influencing factors. The number of factors exceeds the analytical possibilities. No useful model explains the relationship between an impulse and how the situation will change, how the system will react. Experimentation – trial and error – and a step-by-step approach can reduce uncertainties in complex situations. This enables you to make far-reaching decisions more seriously.

In complex situations, there is typically no clear-cut correct solution, no right decision. Complex situations are characterised, for example, by the fact that the majority of experts disagree. Agile Strategy Development is most effective in complex situations with a high degree of uncertainty: by incorporating a wealth of experience and knowledge, gaining knowledge in small steps, and reacting quickly to change, you can master complex situations.

Typical Examples of Experimentation

Developing new markets is a typically complex issue. Developing markets means creating or awakening needs that do not exist. This applies to the department store rental service in Chapters 1 and 2, the 3D agency in Chapter 4, and the pressure sensor company in Chapter 5. These markets are too small, too new, or too specialised to provide reliable data. There is also no law that can reliably predict product sales.

The strategic option »Product development« usually raises complex questions. Either the technical feasibility cannot be estimated without experimentation, or the acceptance of the service by users is uncertain. New technical possibilities change the needs of users, which in turn influences the devel-

opment of devices. Competitors react, which changes demand, expectations, and desires. Both of these largely apply to the new device for analysing carbon from Chapter 1, for example: no comparable device has ever been developed before, and no one can know whether customers will be found for the device. Iterative trial and error is the order of the day.
In the example of the department store from Chapters 1 and 2, it is uncertain what effect «borrowing instead of buying« will have on the company. How will the positioning change? How will the perception of the brand change? Will new services be created? Do manufacturers produce goods differently?

Complex and Complicated Challenges

Your world will hardly be black and white – only complex or only complicated. Mixed forms are typical and relevant to strategizing. The Strategility Framework therefore considers both forms of decision-making – analysis and experimentation – to be valuable. The framework helps you to select the most effective strategy for decision-making depending on the situation.

8.2 Building Agreement in the Face of Uncertainty

As a leader, you are not only required to make smart decisions, but also to reach agreement in the face of uncertainty. Disagreement in the team or among the workforce will torpedo any decision, no matter how wise.

In an organisation, the decision-making process (see Figure 8.2) takes place in the tension between disagreement and uncertainty (see Stacey 2007). The evidence-based approach to agile strategizing moves your organisation away from arbitrary decisions towards fact-based decisions.

Complex situations and times of uncertainty are characterised by a lack of consensus and certainty. The Covid-19 pandemic demonstrated this tension very clearly: on the one hand, there were facts such as infection rates and disease progression and models of virus spread. On the other hand, there was disagreement about the choice of data collection methods or virological models, and about the objectives: personal freedom vs. healthcare system burden vs. overall economic damage.

Below you will find an explanation of the three corner positions: «disagreement with certainty«, «agreement with uncertainty«, and «disagreement and uncertainty«.

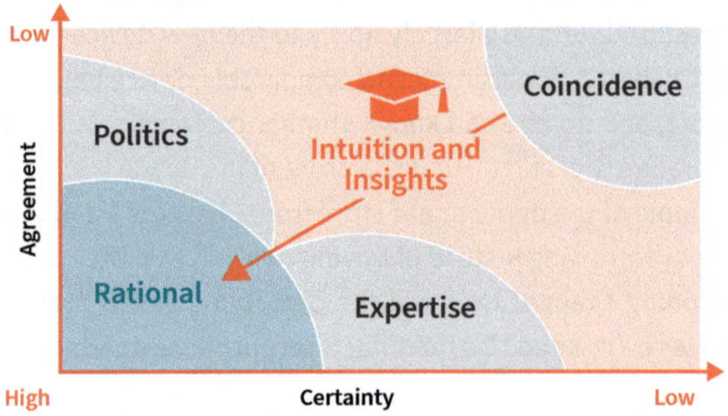

Fig. 8.2: Decision-making in the area of conflict between disagreement and uncertainty according to Douglas Stacey

Low Agreement (Disagreement) with High Certainty
Despite a clear evidence, there may be disagreement regarding decision-making options. Different goals of the parties involved are typical in organisations. On the one hand, these may be personal goals such as career prospects, prestige, power, influence, income, bonuses, etc. On the other hand, individual departments may pursue different goals, and disagreements arise from these conflicting objectives. A typical example is the conflict between cost leadership in a declining product market segment on the one hand and the need for innovation in an emerging market on the other.

Agreement and decision-making are determined by politics. As a manager, you need to have a strong sense of compromise. The participative nature of Strategility can support political decision-making by involving those affected. The step-by-step approach and a learning culture based on insights can also help to achieve consensus.

High Agreement with Low Certainty (Uncertainty)
Agreement on the decision criteria or objectives and on the facts or the chosen method is not always sufficient to decide on options for action. Even when there is consensus on the overall approach, the specific effects may still be uncertain. A simple decision is not obvious.

Decision-making is based on judgement. As a manager, you rely on your experience, intuition or the judgement of knowledgeable people.

The analytical methods and tools of conventional strategizing are most effective when there is agreement – on objectives and methods – but uncertainty about the facts. In such cases, analysis and clarifications can create certainty and enable rational decision-making.

Low Agreement (Disagreement) and Low Certainty (Uncertainty)
Disagreement and uncertainty are the playing field of agile strategizing.

Insights are created in small steps. Uncertainty gradually becomes certainty. Intuitive decisions play an important role in this learning process. They can be made quickly and easily. Tools such as affect balances (see Chapter 8.7) or relative estimates based on the Fibonacci series (see Chapter 2.5) help both to rely on a collective gut feeling and to make intuition tangible and transparent.

Involving stakeholders in strategizing at an early stage can help mitigate conflicts of interest early on and build consensus regarding the approach, models, and procedures so that rational decisions can be made once the facts are available. Decisions made by consultation or by consent (see Chapter 8.7) can also have a unifying effect.

8.3 Hypothesis-Based Approach

The success of strategizing lies in overcoming the strategic challenge. To do this, you must, among other things, turn uncertainty in assumptions into certainty. For complex decision-making situations, it is best to choose an insight-driven method such as experimenting with hypotheses. Figure 8.3 describes such an approach:

1. Have a clear **objective:** Mastering the strategic challenge with favourable decisions
2. Identify the momentous **assumptions** of ignorance, beliefs, and uncertainty
3. Derive **hypotheses**
4. Confirm (validate) or reject (falsify) hypotheses by means of **tests** or experiments
5. Gain **insights** and create a basis for decision-making
6. Make **decisions**

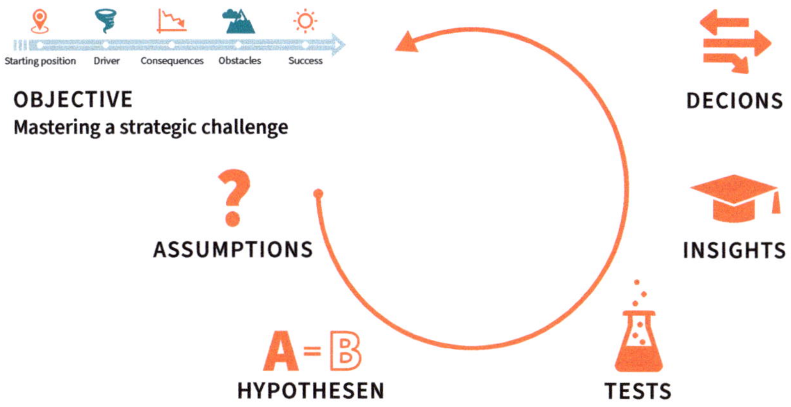

Fig. 8.3: Hypothesis-based approach

8.4 Make Assumptions

Hypotheses provide insight into assumptions. Most assumptions are undisclosed, based on ignorance or beliefs. You can often identify a large number of hypotheses for a specific question. For a reliable basis for decision-making, the momentous assumptions need to be confirmed (verified) or rejected (falsified). Whether an assumption is confirmed or not is in both cases a gain in insights and a success. A common pitfall in practice is that only confirmed assumptions are considered a success.

Another pitfall is that implicit assumptions remain hidden and unverified. Engineers, for example, are typically very enthusiastic about their designs and inventions. That anyone else will be interested in this marvel is a hidden assumption. This phenomenon is very common and is known in the context of product development as «solution looking for a problem« (need).

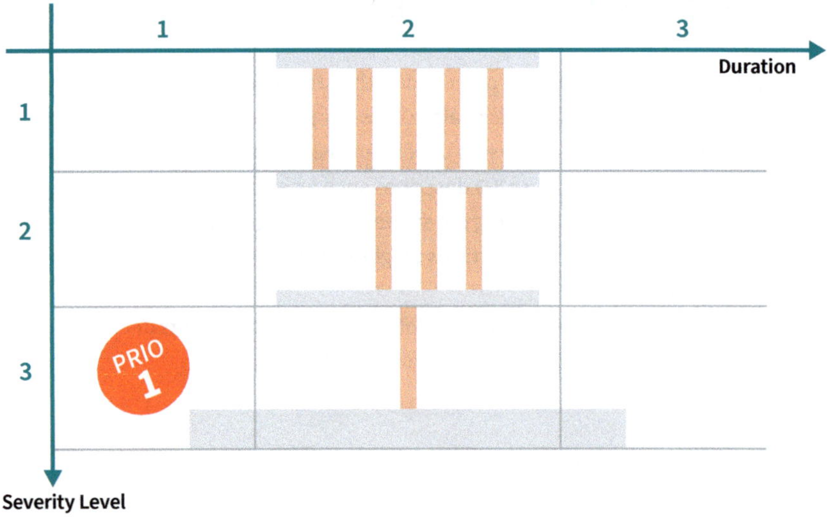

Fig. 8.4: Finding and prioritising the momentous assumptions with the extended risk tower

The extended risk tower (see Figure 8.4) is used to make assumptions visible and prioritise them for clarification.

The work to be undertaken is prioritised in the «Iteration Review & Planning« event (see Chapter 2.7). The momentous assumptions with high uncertainty and high error costs are prioritised in strategizing.

Assumptions can be tested analytically or experimentally/empirically. Analytical methods are suitable for complicated issues that can be answered with expertise and experience. Experimental or empirical methods are appropriate for complex issues. In the following, we discuss experimental/empirical work with hypotheses and tests.

Practical Tip

Workshop: Prioritise Assumptions

1. **Preparation:** Illustration of the risk tower on a pin board (physically in a room or online with an electronic tool)
2. **Introduction:** Explain the strategic challenge and the specific question, for example: Increase company value by expanding into Asia
3. **Brainstorming individually:** The participants collect possible assumptions and beliefs on sticky notes for themselves.
4. **Organise individually:** Participants organise their sticky notes according to the severity of the consequences of an unfortunate impact.
5. **Output:** Each person has their own opinion on the uncertainties.
6. **Merge together (vertically):** In turn, the participants each place a sticky note on the risk tower.
   a) Each participant chooses the sticky note with the highest remaining severity.
   b) When the sticky notes are pinned on the canvas from bottom to top, they should be rearranged if necessary in order to make a new proposal. This rearranging continues until agreement is reached or notes are added to the backlog for detailed clarification.
   c) Similar or comparable assumptions are consolidated.

   Output: Shared view of the momentous assumptions
7. **Estimate the time of impact (horizontal):** The notes of the same severity are lined up from left (earliest impact time) to right (latest impact time).
   Result: The assumptions are prioritised.
8. **Measures:** The assumptions in the field at the bottom left with severity 3 and duration 1 need to be addressed first. To minimise your risk, reduce the duration and/or reduce the severity of the impact.

8.5 Formulate Hypotheses

The aim of empirical work is to convert the uncertainty of assumptions into certainty. Certainty is gained from the results of experiments and tests. Hypotheses are formulated to achieve certainty. A hypothesis is an assertion about the relationship between two factors. This relationship can be formulated using the following model sentences:

- **If ..., then ...**
  - ***If*** *we manage to reach 100 interested parties,*
  - ***then there is*** *a market for our product.*
- **The ..., the ...**
  - ***The*** *more suppliers enter an emerging market,*
  - ***the*** *higher the growth potential.*

8

Several hypotheses can be formulated for an assumption. The formulated hypotheses are organised according to the impact of the assumptions in the Strategility Board. The order of the hypotheses in the backlog results in the prioritisation. The top hypotheses are the most valuable, as their confirmation or refutation provides the greatest certainty in the further course of the project. For the fashion house OVS from the example in Chapter 2.1, a list of hypotheses weighted by delay costs could look as follows:

1. The OVS clothing range, which is successful in Italy, is also popular in Switzerland.
2. Existing Charles Vögele customers will also buy clothes from OVS.
3. The Charles Vögele range can be completely replaced by the OVS range.
4. Existing customers are more loyal to the location than to the brand or the product range.
5. The existing locations of Charles Vögele shops are also attractive for targeted new customers.

8.6 Carrying out Experiments and Tests

Experiments and tests are used to confirm or falsify hypotheses. The focus is on learning success and gaining knowledge. Although confirmations are preferred, every experiment or test with a clear output is a success.

To test a hypothesis, there are often a variety of possible tests and experiments. They typically differ in:

- Effort: Duration and cost
- Significance: the reliability of the outputs and the knowledge gained
- Method: qualitative or quantitative results

The step-by-step approach of iterations in agile Strategy Development is intended to minimise the risk of taking the wrong path. The later a wrong path is identified, the higher the cost of error. For this reason, an intelligent sequence of experiments and tests should be chosen (see Figure 8.5). To become faster and more manoeuvrable, small steps, short iterations with decision-making options, are important. To this end, experiments and tests are divided into three phases:

1. **Rapid feedback before quantitative results**
   Simple experiments with qualitative results can be carried out in a short time, which means that initial findings can be obtained in a short time with little effort.
2. **Low costs of error before high accuracy of outputs**
   In the second phase, the focus is on the affordable loss: how much money, time, or other resources can you afford to lose if you are wrong?
3. **Robust results**
   Once phase 1 and phase 2 have created sufficient certainty, the big experiments and tests can be undertaken. These expensive, momentous experiments yield the most robust and accurate results.

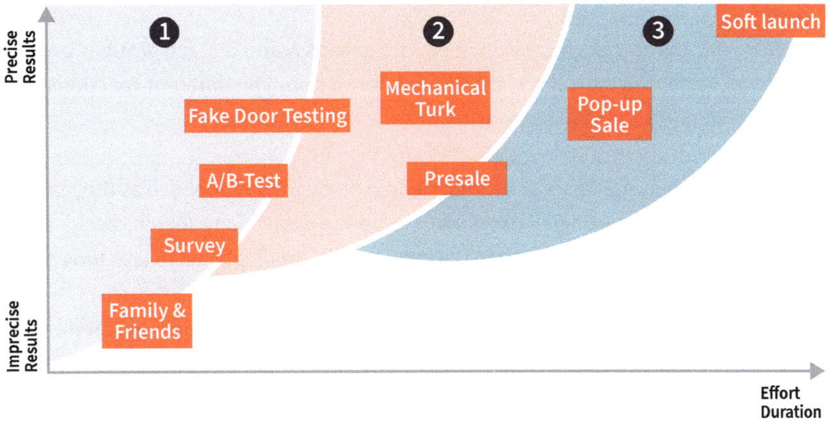

Fig. 8.5: Prioritising experiments

Below we present some examples of experiments and tests.

Friends & Family

The Friends & Family test is the little brother of the survey. The Friends & Family test collects opinions, reactions and feedback from your personal network. Because of the personal reference, the people are immediately accessible, but not very representative. For this reason, the focus is on the dialogue along a few questions with a qualitative assessment.

- **Preparation:** Elevator pitch of the question plus writing down a few questions
- **Realisation:** 10 to 20 face-to-face interviews by telephone, videoconference or on-site
- **Result:** Qualitative analysis of interview notes
- **Duration/effort:** a few hours
- **Note:** Friends & Family should not be taken literally, but symbolically for a good, personal relationship. The ideal counterpart for this test is potentially affected or interested and must be willing to do a favour and be able to keep the content confidential. This type of informal survey is very suitable for companies with small groups of customers for whom surveys are not worthwhile.

The Survey

Surveys are popular and tried and tested. Properly designed surveys can be analysed quantitatively very well. Due to the standardisation, individual statements can be clarified very well through surveys. On the other hand, this also limits the findings and conclusions that can be drawn from the question. Surveys are therefore most valuable for clarifying detailed questions and narrowing down search fields. Surveys complement experiments such as the A/B test or the landing page very well.

8

A/B test

In the A/B test, the target group is divided into groups A and B. Each group is provided with its own variant of a product, service or information. The different reactions of the two groups can be used to draw conclusions and test hypotheses.

Experiments such as A/B tests show that people do not always behave as they say they will. Surveys show that Google[10] users want to be shown as many search results as possible. In A/B tests, however, shorter result pages that are displayed in less time proved to be more popular.

- **Prerequisite:** A large target group is required for statistically relevant statements. It must be easy to provide two variants of the offer and also to remove and dismantle them again. Furthermore, the behaviour of the target group must be easy to record and evaluate.
- **Implementation:** Typical applications are web applications, advertising messages or electronic adverts. A/B tests for websites or web applications typically run for one to two weeks with a target group of around 10,000 people.
- **Duration/effort:** If the technical requirements for A/B tests are met, you can expect two to four weeks of preparation and an additional two weeks to analyse and process the results.
- **Result:** The hypothesis is confirmed or falsified by quantitative results.

Fake Door Testing

Fake door testing is used to simulate products or services that do not actually exist. The aim is to test the attractiveness of a potential offer. Fake door testing is worthwhile when the costs of disappointed potential customers are lower than the costs of effectively delivering the offer.

Experimenting with fake door testing is very popular for digital products (such as fake online shops), as they are easy to create and customise, and the user behaviour can be tracked and evaluated very accurately using telemetry data. Typical examples are landing pages or buttons that simulate an offer. Visitors can then only register for further information.

- **Preparation:** For a fake door experiment, you need to plan not only the provision of the fake door, but also the evaluation and cancellation of the experiment. Managing expectations and then disappointing them is a balancing act.
- **Duration/effort:** The duration of the experiment depends on the size of the target group and the ability to attract attention. Experiments with landing pages are therefore often combined with electronic adverts. Good results can be expected in two to eight weeks.
- **Result:** By faking a product or service, qualitative and quantitative results are generated in terms of genuine customer interest.

10 Google I/O ’08 Keynote by Marissa Mayer.

A state-owned natural gas supplier was trying to enter the electric vehicle charging business for a more sustainable future. Within two months, a small team designed their own brand, created a landing page with various offers, and launched an advertising campaign. Despite a positive response, the project was cancelled for political reasons. The fake door technology saved the company from having to write-off the cost of delivering the real product.

Mechanical Turk

The term dates back to a mechanical chess automaton from the 18th century. The chess machine contained a human chess player who controlled a puppet dressed in traditional Turkish costume. When experimenting with the Mechanical Turk, individual components of a product are provided with a workaround solution. The inventor was unable to develop a chess machine and switched to using a hidden person.

- **Implementation:** In experiments with a Mechanical Turk, long-term investments are postponed by short-term operating costs.
- **Result:** Time to market can be gained or the risk of a bad investment reduced. If the chosen path proves to be viable, the Mechanical Turk will be replaced by a sustainable solution over time.

In our example of the department store, a large part of the logistics processes were provided as a Mechanical Turk (fake process). The products ordered for rental from the web shop were regularly purchased, packed, and shipped by employees at the checkout in the store. The return of products was not provided. Customer service was only instructed to contact the project team in the event of a return. The project team built the website with a functioning online rental service within two weeks.

Advance Booking

Pre-sale is a variation or an extension of fake door, where the buyer and seller enter into contractual obligations during the pre-sale (or reservation). Aircraft manufacturers only go into production after a large number of reservations have been made. Tesla offered vehicles for pre-sale years before the start of production.

- **Use:** Assumptions about sales opportunities and price structures can be tested.
- **Result:** Production or the start of production can be financed with the secured presales and the early revenues.

Pop-up Store

A pop-up store offers services or goods for a limited period of time. The store appears suddenly and then disappears again. For this purpose, unused sites are usually used on a temporary basis.

- **Use:** Testing markets and target groups for new products, locations, sites, places or countries
- **Advantage:** Low operating costs and no long-term commitment by using empty space for a limited period of time

8

- **Results:** Experiences and insights of a qualitative and quantitative nature along the entire customer interaction chain

Soft launch

With a soft launch a product or service is introduced «quietly«, without announcement.

- **Use:** This allows you to limit or avoid quality problems when launching products. You can either improve the product itself or prepare to scale the business.
- **Result:** The knowledge gained from the use of the product is used to optimise and improve the provision or the service itself.

8.7 Gaining Insights and Making Decisions

Successful tests and experiments provide new insights and create certainty. The results confirm or falsify the underlying hypotheses and assumptions. From the point of view of gaining knowledge, a disproved assumption or a rejected hypothesis is just as valuable as a confirmation.

Insights are gained not only from the results – confirmed or falsified – but also from the analyses. The findings are used to refine backlogs, plans and the next steps.

The decision is: continue or pivot. Pivoting means leaving the chosen path and taking a new direction. In start-up product development, these pivot events are specifically provoked in order to sharpen and improve the product or business.

Decisions in Hypothesis-Based Work

- **Assumptions confirmed – move on:** The chosen assumption proves to be feasible. With further experiments, the less risky hypotheses can now be tackled with tests and experiments.
- **Assumption falsified – dead end:** The outputs and insights show that the assumptions made are invalid. Pursuing the path further would lead to a dead end or have serious consequences. Returning to square one may mean:
  - The strategic option cannot be realised with the developed strategy. Either the strategy or the strategic option is discarded and an alternative is developed.
  - An individual or group of strategy elements will not be able to develop their full effect. Alternatives need to be developed.
  - The implementation of the strategy is not achieving the desired impact. The Strategy Implementation needs to be corrected or the strategy element in question needs to be revised.
- **Assumption refined – deepen:** The findings show that the assumptions are at least partially correct. However, a course correction for the further course is recommended. For example: The potential customers belong to a different target group. Subsequent experiments and tests refine the findings and consolidate the path.

Practical Tip

Decide Quickly with the Affect Balance

The concept of «fast thinking, slow thinking« (Kahneman 2011) divides our decisions into fast, unconscious and intuitive on the one hand and slow, conscious and rational on the other. As a Strategility Master, the affect balance (Storch 2000) offers you a simple technique to make balanced decisions in just a few minutes:

1. Participants are prepared to respond spontaneously to a question.
2. The Strategility Master formulates the decisive question.
3. Participants record their positive and negative spontaneous reactions on two scales: approval and disapproval.
4. The participants explore the reason for the spontaneous reaction.
5. The decision is made – as an individual decision, by consensus, or consent.

Decision-Making Process

Teams are superior to individuals when it comes to solving complex problems. Agile methods therefore rely on teamwork and the inclusion of different opinions and specialisms. Involving many people (including the entire workforce) in strategizing can be valuable. However, this does not mean that everyone has a say. This is partly because only a few – management, board, shareholders – have to bear the consequences of strategic decisions, and partly because in urgent cases individuals can make decisions more quickly than committees or a larger team. There is a wide range of decision-making procedures, from individual decisions taken by the boss alone to a vote of the entire workforce. Transparency about the processes involved in the work on the Strategility Board, the results of the work with the hypothesis board and about the joint review and planning work means that the basis for decisions is open. Creating a basis for decision-making is one of the main tasks of the Strategy Exploration Team, and it is therefore involved in the decision-making process.

It is not only the number of people involved in a decision that can have a delaying effect, but also great uncertainty or a poor basis for decision-making. Strategility's risk-based, empirical approach allows you to create a basis for favourable decisions. The step-by-step approach also cushions the impact of unfavourable decisions. This approach leads to a high level of decision-making capability and contributes more to rapid decision-making than a single decision.

Proven decision-making processes in Strategility are

- **Consultative individual decision**
  Through transparency and cooperation, individual decisions are based on consultation. The individual decision allows the Enabler to fulfil their responsibility and lead the company strategically. When an individual fulfils the role of Enabler, for example in owner-managed SMEs, this is a common form.

8

- **Decision by consent**
  A decision by consent among the participants is reached in the absence of serious objections. Participants are more involved than in a pure consultation process. The participants support the decision.
- **Decision by consensus**
  A decision by consensus among the parties involved is reached by agreement. It must be decided whether unanimity is required to reach a consensus or a majority is sufficient.

Agreeing to a proposal requires a higher level of conviction than the absence of a serious objection. An objection is serious if it can be shown that a proposal poses a serious risk. Agile methods and responsive organisations are based on decision making by consent.

> **Practical Tip**
> **Moderated Decision by Consent**
> The decision is made by consent. The Strategility Master moderates the decision-making process as follows:
> 1. Formulate request
> 2. Clarify questions of understanding
> 3. Raise objections
> 4. Propose a solution
> 5. Start again with a new request
>
> If it is not possible to overcome all objections in a few rounds, it is advisable to postpone the decision. In order to reach a decision, tasks are distributed to create a better basis for decision-making.
> In the event of disagreement, consensus and consent often consolidate the status quo. For rapid decision-making in urgent cases, we recommend the consultative individual decision.

The extent of participation depends on several factors:
- the urgency
- the need to involve and include the workforce
- the structure and culture of the organisation

Table 8.1 presents typical decision-making procedures for key elements of Strategility: from low to selective to comprehensive participation and from rapid to rather lengthy decision-making.

	Low participation	Selective participation	Comprehensive participation
Mission statement	Consent strategy team	Consent workforce	Staff consensus
Strategic challenge	Individual decision enabler	Consent strategy team	Consent workforce
Strategic option	Individual decision enabler	Consent strategy team	Consent workforce
Strategy	Consent enabler and strategy owner	Consensus strategy team	Consent workforce
Strategy elements	Individual decision strategy owner	Consensus strategy team	Consent specialist area
Priorities in the strategy backlog	Individual decision enabler	Consensus strategy team	Consensus strategy team
Initiative implemen-tation	Individual decision strategy owner	Consensus strategy team	Consent specialist area
Implementation control	Individual decision strategy owner	Consensus strategy team	Consent specialist area

Tab 8.1: Decision-making procedures in strategizing – from quick decisions with low participation to lengthy decisions with extensive participation

8.8 Working with the Hypothesis Test Canvas

The hypothesis test canvas (see Figure 8.6) helps you keep an overview of the work with hypotheses and tests throughout strategizing. The canvas provides orientation across strategy teams, ensuring a common direction. A common direction is necessary so that individuals and teams can carry out detailed work autonomously without losing sight of the big picture. Each canvas represents a single or a few related questions. The canvas accompanies an Exploration Team through several sprints and one to three Strategility iterations.

The canvas consists of:

1. The orientation parameters from the Strategy Orientation: mission, vision, strategic challenge (strategic option)
2. The assumptions to be tested – including status – from the risk tower
3. The hypotheses – including status – on the assumptions
4. The formulated experiments and tests – including measured variables and status – for the hypotheses
5. The test results and findings as well as further observations
6. Valuable information and observations beyond the qualitative and quantitative results typically obtained from experiments and tests with direct interactions.

The canvas is not static, but is continually updated until the assumptions to be tested are finally confirmed or rejected. The content of the canvas is developed during the various events of the Exploration Team. You can also use the hypothesis and test canvas from Chapter 4 for this purpose. The content is developed by the members of the Exploration Team. The Hypothesis Owner is responsible for the results. The role of the Strategility Master is to guide the team in the application of the canvas and in empirical work with hypotheses and tests. They are also responsible for moderating the various events and workshops.

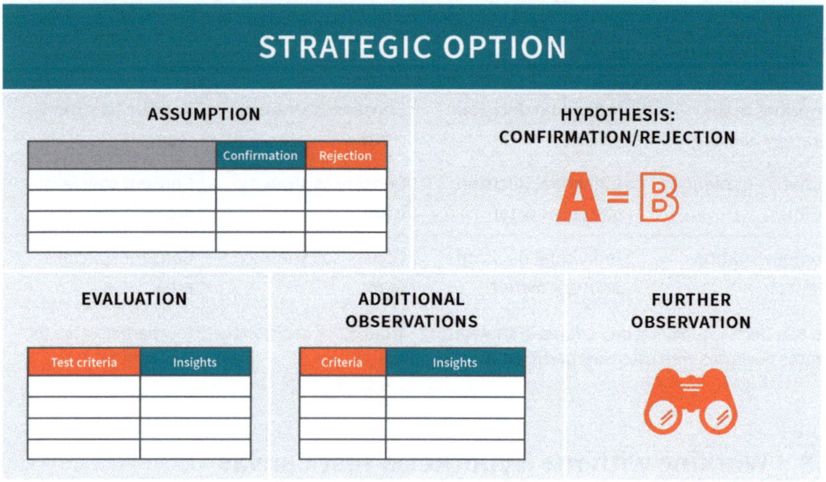

Fig. 8.6: The hypothesis test canvas

CHAPTER 9

Conclusion

9 Conclusion

In this toolbox, you have learned about processes, tools and roles for combining long-term strategizing and short-term, responsive agility. Finally, we would like to summarise key messages and selected recommendations according to the chapter structure:

1. The iterative, evidence-based, and impact-oriented Strategility method is particularly suitable for **strategy questions with a high degree of uncertainty.** Choose analytical decision-making for questions for which you can gather sufficient data or experience. It is probably faster and less time-consuming. On the other hand, you should always use the element of **continuous information and knowledge flow** between Strategy Orientation, Strategy Development, and Strategy Implementation in order to quickly identify and react to strategic changes.
2. A key element of Strategility management is the Review & Planning event, where all those involved in the strategy discuss the outputs and impact of strategic decisions and agree on further decision-making steps. As a result, strategic decisions can remain open for a relatively long time – which is not always easy to endure. We have also observed a tendency for Strategy Exploration Teams to identify with the strategy options they are exploring. The Strategy Owner should weigh up the content of the upcoming strategic decisions with an explicit **factual orientation.**
3. In the Strategy Orientation, you define the long and medium-term development of your organisation. Two aspects need to be clarified:
   a) What are the owners' plans and what is your future mission?
   b) What is the strategic challenge for your organisation?
   These two questions are important orientation parameters for all your further work, but they are often not answered explicitly. **The ownership strategy and mission** are the guiding principles for your entire economic and content-related orientation and the **strategic challenge** is the overarching evaluation criterion for all assumptions and decisions. All strategic activities should contribute to mastering the strategic challenge.
4. In agile Strategy Development, you make assumptions about strategy elements explicit and verifiable. Many of our practice partners find the **Strategility Hypothesis Canvas and the Strategility Test Canvas** useful for structuring and clarifying this step. The canvas builds a bridge between the strategic planning approach and the agile testing approach and helps to constructively combine the two approaches and consciously reflect on intuitive decisions.
5. Agile Strategy Implementation contributes significantly to the quality of Strategy Development. The **impact of Strategy Implementation provides you with information about your strategic assumptions**: Are they proving to be true, and what significance do they have for overcoming the strategic challenge? Carry **out an ongoing evaluation of Strategy Development during Strategy Implementation** by analysing the impact of implementation as well as the implementation impacts. En-

9

ablers and Strategy Owners are required to be prepared to change direction during Strategy Implementation as part of their responsibility for effectiveness.

6. For agile Strategy Implementation, it is helpful to differentiate between key roles, such as the role of the Enabler, who is responsible for the strategic direction, or the role of the Hypothesis Owner, who is responsible for reviewing the strategic assumptions. Even if you combine these roles, you should emphasise the different tasks and responsibilities. With some of our practice partners, we have seen that Enablers and Strategy Owners were ambitious and enthusiastic about driving a strategic idea forward. Deliberately handing **over strategic ideas to the Hypothesis Owner** for critically evaluation enables reflective impulses and evidence-based discussions.
7. In agile Strategy Development, your corporate culture is required to combine the different basic attitudes of agility and strategic management. Based on our experience, we recommend that you become aware of your prevailing corporate culture, make the importance of both basic attitudes explicit and **address** any **cultural gaps or tensions through processes** rather than attempting a comprehensive cultural change. This is where the Strategility Master plays an important role in process design: consciously organise events with agile and strategically dominated parts, and design the discourses as intervals of listening, questioning, and assumption-based decisions.
8. To enable you to drive your decisions effectively and efficiently with the help of experiments and tests, we recommend that you consciously **prioritise**:
   a) **The most momentous assumptions**, the refutation of which will knock out an entire strategy or strategic option
   b) **The most insightful hypotheses**, the testing of which will provide the most meaningful information
   c) **Simple tests and experiments** that will provide you with initial information in a short time and with as little effort as possible

   And, of course, we would like to remind you of the reassuring phrase: »**Every experiment and every test with a clear output is a success**« because you will know more than you did before, even if it is the knowledge that your strategic assumption is wrong.

We are convinced that this toolbox has given you a comprehensive insight into a wide range of experiences in agile Strategy Development. We now look forward to hearing about your experiences and welcome your feedback and questions:

- *Thomas Haas:* Are you dissatisfied with the speed and progress of strategizing? Would you like more courage, more co-creation, more collaboration or even an agile organisation? As an entrepreneur, consultant and lecturer, these are precisely the questions that have been driving me for over 20 years, and I look forward to your feedback and your very personal challenge.
- *Bernhard Kruschitz:* Using energy in the right place and getting everyone involved to work together are the most important entrepreneurial challenges. I am convinced that the Strategility Toolbox with its underlying agile principles provides a good ba-

sis for mastering these challenges. Of course, questions will arise during application. Experience has shown that it is easy to complicate something, but challenging to simplify it effectively. I look forward to your questions and will be happy to support you with simple answers.

- *Kerstin Pichel:* Your individual design of the Strategility process inspires me for my further research – after all, as a lecturer in strategic management, I have been dealing with the effective design of strategy processes for years. Strategy Orientation variables and their interplay are part of my basic training and passion – especially in strategy consulting. I look forward to your feedback and your questions about the strategy process and the interplay of strategic variables.

Bibliography

Anderson, D. J. (2011): Kanban: Evolutionary change management for IT organisations. Heidelberg.

Ansoff, H. I. (1965): Corporate strategy: An analytic approach to business policy for growth and expansion. New York.

Belbin, M. (2004): Management Teams: Why They Succeed or Fail. Oxford.

Hambrick, D. C./Frederickson, J. (2005): Are You Sure You Have a Business Strategy. Academy of Management Executive, 51.

Johnson, G./Whittington, R./Scholes, K./Angwin, D./Regnér, P. (2018): Strategic Management. 11th edition, Hallbergmoos.

Kahneman, D. (2011): Thinking, fast and slow. New York.

Koryak, O./Mole, K./Lockett, A./Hayton, J./Nicolaou, N. (2018): Disentangling the antecedents of ambidexterity: Exploration and exploitation. In: Research Policy 47(2), pp. 413–427.

Porter, M. (2008): Competitive forces – a new look. In: Harvard Business Manager 30(5), pp. 20–26.

Reinertsen, D. G. (2002): An Overview of Cost-of-Delay Analysis: Calculating Project Decision Rules. In: The Journal of Cost Analysis & Management 4, pp. 8–24.

Rüegg-Stürm, J./Grand, S. (2020): The St. Gallen Management Model. 2nd edition, Bern.

Schneider, W. (1994): The Reengineering Alternative: A plan for making your current culture work. New York.

Snowden, D. (2021): Managing complexity (and chaos) in times of crisis. A field guide for decision makers inspired by the Cynefin framework. Luxembourg: Publications Office of the European Union.

Stacey, R. D. (2007): Strategic management and organisational dynamics: The challenge of complexity to ways of thinking about organisations. Harlow.

Stadler, C./Hautz, J./Matzler, K./von den Eichen, S. (2021): Open Strategy: Mastering Disruption from Outside the C-Suite. Cambrigde/London.

Storch, M. (2000): The Zurich Resource Model CRM. In: Contributions to teacher training, 18(3), pp. 307–323.

Thode, S./Wistuba, L. (2019): $SWOT^3$ as a method of agile strategy development. In: Dahm, M./Thode, S. (eds.): Strategy and transformation in the digital age. FOM Edition. Wiesbaden, pp. 22–38.

The team of authors

Kerstin Pichel

Prof. Dr Kerstin Pichel is an energetic and engaging lecturer in strategic management at the Institute for Organizational Viability at the Zurich University of Applied Sciences. With over 15 years of experience researching, advising, and publishing on current challenges in strategy processes, she has developed a highly practical and applicable approach to her scientific findings. This makes her research highly accessible and inspiring, providing a solid foundation for others to engage with strategic management. The insights and experiences derived from numerous case studies from diverse sectors, including industry, services, and non-profit organisations, provide a compelling illustration of the recommendations put forth for contemporary strategy processes. Ms Pichel derives sustenance for her enthusiasm from a variety of activities in nature and from deeply connected relationships and friendships.

Thomas Haas

Thomas Haas is thrilled about the change and the possibilities of today. With his various activities as an entrepreneur, consultant, and lecturer, he helps companies and organisations realign for sustainable growth: digital, agile, human centred. He graduated in business informatics at the University of Zurich and worked as a software engineer in early innovation and digitisation projects at the turn of the millennium. With the dotcom crash, he focused on corporate development and transformation by digital means and has been helping companies and organisations to thrive ever since! He is particularly passionate about creating growth-capable organisations and motivating working environments. Thomas Haas is an inspiring keynote speaker at conferences and in professional organisations, as well as an endurance athlete in nature. He is the proud father of grown-up children and sometimes just likes to be at home.

Bernhard Kruschitz

Bernhard Kruschitz is an entrepreneur, consultant, trainer and speaker who is known for his engaging seminars and workshops, which have now been attended by over 20,000 managers. As an entrepreneur, he knows the ins and outs of running a business. Originally a business informatics specialist, he founded BKI AG, a company specialising in project management, and successfully managed it as owner for over 20 years. He recently handed it over to his successor. As a project management expert, he has developed and introduced project management methods for companies in a wide range of industries, sizes, and cultures, such as the HERMES 5 project management method of the Swiss Federal Administration. As a strategy consultant, Bernhard Kruschitz brings energy and structure to agile, evidence-based Strategy Development and Implementation. At the centre of his actions are people and their successful cooperation in change.

Index

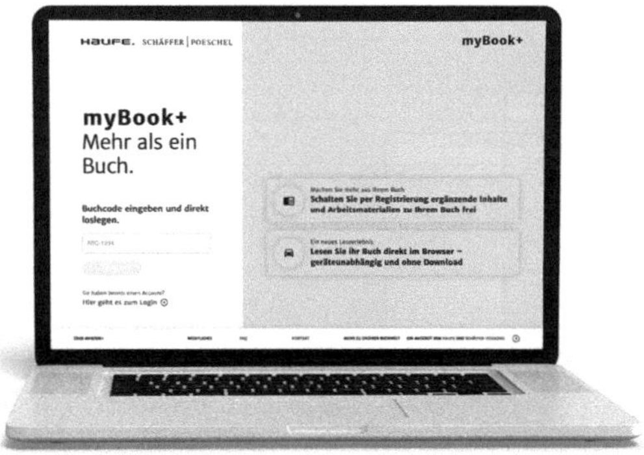
HAUFE. SCHÄFFER | POESCHEL
myBook+
myBook+
Mehr als ein
Buch.
Buchcode eingeben und direkt
loslegen.